Insects in the City

An archaeoentomological perspective on London's past

David Smith

BAR British Series 561
2012

Published in 2016 by
BAR Publishing, Oxford

BAR British Series 561

Insects in the City

ISBN 978 1 4073 0986 6

BAR Publishing is the trading name of British Archaeological Reports (Oxford) Ltd. British Archaeological Reports was first incorporated in 1974 to publish the BAR Series, International and British. In 1992 Hadrian Books Ltd became part of the BAR group. This volume was originally published by Archaeopress in conjunction with British Archaeological Reports (Oxford) Ltd / Hadrian Books Ltd, the Series principal publisher, in 2012. This present volume is published by BAR Publishing, 2016.

Printed in England

BAR titles are available from:

BAR Publishing
122 Banbury Rd, Oxford, OX2 7BP, UK
EMAIL info@barpublishing.com
PHONE +44 (0)1865 310431
FAX +44 (0)1865 316916
www.barpublishing.com

Table of contents

List of Figures

Preface

Books on archaeological insects are rare, which cynics might suggest is hardly surprising: insect remains hardly leap to the attention of the average excavator. But insect remains from archaeological sites can tell us an astonishing amount about the past. This ranges from prosaic lists of which species were present (and even these tell important stories about trade and social change), via intimate details of the parasitological state of peoples' privates, to socially and economically significant reconstructions of the environment and climate.

However, many insects are unfamiliar to most people, and the methods used to glean information from their fossils can be complex. What David Smith has done in this short book is to make us feel much more familiar with the creatures themselves, and subtly entwine descriptions of site results, explanations of methodology, and outlines of the conclusions in a readable way. As we read, we learn why weeks and months of peering down microscopes struggling to extract and identify tiny bits of insect are worthwhile. More than that, we can understand how the (often unexciting) details of remains for single sites can be woven together into bigger stories.

The results from London, with their long time span and geographical range, present an excellent basis for an accessible account of this kind. This book will bring them to a wider audience, which is commendable. But even more importantly, this book should serve to convince more archaeologists (and their funders) that bioarchaeology in general, and work on insect remains in particular, is worthwhile and deserves a fair cut of the project budget.

Harry Kenward, Department of Archaeology, University of York, former Director Environmental Archaeology Unit, University of York.

Acknowledgements

First, I would particularly like to thank the Museum of London Archaeology Service for giving permission to publish this book. I am grateful to James Rackham, John Giorgi, Anne Davis and Jane Siddell for having the initial confidence in me when they gave me the first round of work on behalf of MoLAS in the early 1990s. They all have continued to present me with material and opportunities over the next decades and I am grateful for their continued support and collaboration. More recently, Pete Rowsome (also MoLAS) has been particularly encouraging.

First drafts of this book have been commented on and corrected by Harry Kenward, Nicki Whitehouse, Jane Siddell and Pete Rowsome, who all gave very helpful comments and looked on my failings with charity. I also am indebted to Wendy Smith for proof reading various and endless drafts of the text.

The idea of writing this book originally was dreamed up by Prof. Leslie Brubaker at a research review. It's all her fault.

I am grateful to Chris Caseldine, Ralph Fyfe and Hari Hjelle for allowing me to see a pre-publication proof of their 2008 paper in *Vegetation History and Archaeobotany*. Similarly, Prof. Scott Elias allowed me to see a pre-publication copy of his paper in the *Geology Journal*.

The majority of this book was written whilst in receipt of an Arts and Humanities Research Council 'Research Leave' Award (Award no 113058). I am also grateful to my home institution, The University of Birmingham, for providing 'matching leave' extending my sabbatical period to cover the entirety of the 2005/ 2006 academic year, which enabled me to prepare a complete first draft and begin circulation and editing of the second draft.

Figure acknowledgements

Bryony Ryder and Nigel Dodds, to whom I am indebted, prepared Figures 2.1, 3.1, 5.1 and 7.1. Figures 2.3 and 2.4 were prepared by Henry Buglass and were redrawn from Coope and Brophy (1972) and Atkinson *et al.* (1987). Both were published in Gaffney *et al.* 2006 and are reproduced here with permission of the authors.

Figures 5.2, 5.4, 5.5, 7.2, 9.2, 9.3 and 9.4 were drawn by Henry and Louise Buglass. Figures 5.2 and 5.5 were redrawn from Rowsome (2000). Figure 5.4. is redrawn from Bateman 1997. Figures 7.2 was redrawn from Rowsome and Treviel 1998. Figures 9.2 and 9.3 were redrawn from Thomas *et al.* 1997. Figure 9.4 was redrawn from Slone and Malcolm 2004.

Figure 2.2 is redrawn from Sidell *et al.* 2000 with permission of the authors. Figure 10.1 is reprinted with the permission of the Council of British Archaeology. Figure 10.2 and 10.3 is reprinted with the permission of the publishers of the *Journal of Archaeological Science*.

Figures 11.1, 11.2, 11.3 and 11.4 were prepared using the CANODRAW programme (ter Braak and Smilauer 2002).

CHAPTER 1: INTRODUCTION

GENERAL COMMENTS AND THE STRUCTURE OF THIS BOOK

This book had a quiet and organic genesis. It started life as an eighty page technical summary of my work on the insect faunas from Central London. It was intended to provide a limited example of how to use a range of multivariate statistics on archaeological insect faunas and an attempt to compare this to similar work undertaken from York (Carrott and Kenward 2001). Through chance, misfortune, opportunity and 'random drift' it has ended up developing into a more interesting beast. It now deals with several different strands of information:

1) It describes the insect faunas recovered from a number of geological and archaeological sites in Greater London and the city. These now cover the period between 350,000 years ago up to the end of the 16th century AD.

2) It now includes discussion on the changing landscape of the Thames basin during this time along with the development of the 'living landscape' of the city.

3) The insect faunas are used to examine and highlight a number of important research questions for the geological and archaeological periods concerned. Often these have direct relevance for our understanding of how humans and society functioned at the time.

4) The various ways in which insect remains can be used to interpret the archaeological record are explored. This includes a range of technical and theoretical developments that have happened recently in both archaeoentomology and more generally in environmental archaeology.

5) There also is a summary of the history and archaeology of the Lower Thames basin and the city of London. This is necessarily partial in terms of coverage. For example, I do not discuss the last period of the Roman occupation in detail. Though fascinating in itself, there are no insect faunas from London for this period. I apologise for such selectivity but it is unavoidable in a book about archaeological insects.

I also have chosen not to follow the traditional line with 'scientific' archaeological books such as this. Usually there is a chapter outlining the theoretical basis and the techniques used (sometimes that is all there is), followed by a chapter that gives the history of research in the area of study and raises prominent research issues. Then there is a results chapter followed by a conclusion that shows how the research issues identified have been addressed. I have decided to avoid this rigid structure. I believe that the themes raised above are better blended together to form an interlocking and coherent chronological story.

A HISTORY OF ARCHAEOENTOMOLOGY IN GENERAL AND IN LONDON IN PARTICULAR

Palaeoentomology (the study of insect remains) is a relatively young discipline. The earliest attempts to look at the insects from geological sediments were usually exercises in species spotting rather than environmental interpretation and were limited in their scope. Examples are Blair's work on a number of coastal 'peats' (Blair 1924; 1935) and Bell's (1920; 1922) on a number of Pleistocene deposits in Britain. On the mainland of Europe workers such as Kolbe (1894; 1925), Lomnicki (1894) and Mjöberg (1905; 1915) analysed similar deposits at a number of locations. Scudder (1900) undertook early work in the Eastern United States and Canada, with later work in the U.S. concentrated around the tar pits of southern California, mainly by Pierce (e.g. Pierce 1947; 1953). One problem with this work was that it was slow and laborious. Peat and other sediments had to be split along 'fault' lines and searched by eye under a microscope. This resulted in the recovery of a rather incomplete fauna since only 'spectacular' and large insects were routinely recovered (Coope 1961). The main difficulty, however, was that workers in the field assumed that considerable evolution should have occurred in the intervening geological period and that the majority of the species encountered would be extinct (Buckland and Coope 1991; Elias 1994, 2010). This all lead to the misidentification of many specimens with several 'extinct' species being given rather unusual names. My favourite is *Bembidion damnosum* recorded by Scudder (1900). With so many extinct species it was also thought that such faunas had a limited role in interpretation (Elias 1994, 2010).

Much of this changed in the 1950s. The first key event was the development of 'paraffin flotation'. This processing technique allows the quick recovery of large and representative insect faunas (Coope and Osborne 1968; Buckland and Coope 1991, Rousseau 2011). The second was Coope's early work that established there had been no speciation during the Quaternary (the last 500,000 years) and, therefore, insect remains had great potential for interpretation of the past (Coope 1978; 2004). There have been some recent attempts to suggest that the lack of speciation or change in insect faunas may be limited to Northern Europe with speciation occurring at a faster pace in the tropics (Ashworth 2004). Coope (2004) suggests that this may relate to the rapid patterns of climate change in Northern Europe throughout the period keeping populations diverse and mixed effectively preventing isolation or genetic drift and mutation. Certainly Ponel *et al.* (2003) looked at 110,000 years of deposit and found no speciation in the sequence. A more detailed summary of this current argument can be found in Whitehouse (2006) and Elias (2010).

The earliest attempts to use insect remains in the archaeological record were often the finds of 'noticeable' insect remains in archaeological contexts.

Classic examples include Alfieri (1931) on the pests of stored products found in the tomb of Tutankhamen and Ewing (1924) on the lice recovered from human mummies. Perhaps more systematic was the work of Strobel and Pigorini (1864) on insects found in the Italian 'lake settlements'. Once again the key 'breakthrough' came in the laboratories in the Geology Department at the University of Birmingham in the early and mid 1960s. In addition to working on Quaternary deposits Peter Osborne, Russell Coope, Paul Buckland, Harry Kenward and Maureen Girling all showed an interest in archaeological material. The work produced by this *cadre* clearly demonstrated the validity of looking at insect remains to address archaeological questions. This lead to the development of a number of large projects; such as the analysis of Anglo-Scandinavian deposits at Coppergate, York (Buckland *et al.* 1974; Hall *et al.* 1983) and the work undertaken by Girling on the Somerset Levels trackways (Girling 1977, 1979a, 1980). Though not from the same stable, Mark Robinson (Oxford University) developed and promoted archaeological insect analysis at this time in the Upper Thames Valley (Robinson 1978; 1979). It should be noted that English Heritage supported all the individuals who worked on these projects. In terms of providing initial funding and encouragement for archaeoentomology the role of English Heritage cannot be underestimated. I belong firmly to the next generation. I merely follow in the footsteps of those who went before. Elias (2006, 2010) has outlined the recent spread of archaeoentomology outside of northern Europe, along with some recent trends in the discipline.

With the exception of a number of Quaternary sites studied by Coope (these are discussed in Chapter 2), previous work on archaeological deposits in London is very limited. Maureen Girling looked at a number of faunas from the Southwark excavations in the 1970s (Girling 1979b) and a limited number of 18th century deposits (Girling 1984). Kenward also undertook analysis of the more extensive material from the excavations along the Walbrook in the 1980s (de Moulins 1990). Archaeoentomology in London waited until I started to cast around for work in the early 1990s. I had met John Giorgi of the Museum of London Archaeology Service (MoLAS) when he came to Sheffield to study for an MSc whilst I was doing my PhD research. I wrote to him in 1991 asking if MoLAS had any insect analysis that needed undertaking. This was not altruism: I needed the cash since my PhD funding was finished and I was 'between jobs', so to speak. He raised this with James Rackham, who was then in charge of environmental archaeology at MoLAS, and he seemed interested. I duly went down to London where I was interviewed and shown a few fragments to identify. I seemed to pass the test and went home with some flots left over from the studies of plant remains. Shortly afterwards, I got a job as a technician at the University of Birmingham where Prof. Susan Limbrey encouraged me to do more work like this. Peter Osborne and Russell Coope checked my initial set of identifications from this material. All those named above, who crucially provided help and support with my first 'London contracts' need to be thanked. Without their support at this early stage in my career, I would not have worked on this material or learnt how to do the job, and MoLAS would not have fallen into the habit of sending me their material.

BASIC TECHNIQUES – SAMPLING, PROCESSING AND IDENTIFICATION

Insect remains are found in a wide variety of geological and archaeological deposits (Buckland and Coope 1991, Elias 2010). The key thing is that the deposits have to be desiccated, frozen or waterlogged since they were laid down. Any wetting of desiccated remains or drying of waterlogged material, and any oxidisation or exposure to biological decay, will destroy insect remains. Waterlogging is the most common way insect fragments are preserved in the British Isles. This is quite a rare event. I would estimate that approximately 10% of the archaeological sites in Britain have at least some waterlogged layers.

The sampling and processing of material is straightforward and is covered by in a number of guides (Dobney *et al.* 1992; English Heritage 2002). In terms of urban sites a 10 L sample is often collected from each deposit on site that seems to have waterlogged preservation and is of archaeological significance. This can produce a mountain of samples so a process of selection and assessment is carried out in order to decide which should be worked on. How this selection is attempted, and the potential pitfalls that can occur are discussed in Chapter 8. Back in the lab, the material is paraffin floated. This process was outlined by Coope and Osborne (1968) and refined by Kenward *et al.* (1980). Basically, the sample is sieved over a 0.3mm mesh and the contents of the sieve are retained. Approximately 100ml of paraffin is added to the material retained by the sieve and is blended in, usually by hand. The result looks somewhat like damson jam. The paraffin binds to the surface of insect fragments so that when cold water is added they float and are concentrated on the water surface. With luck most other material such as sand, stones, plant remains and other 'residue' remain at the bottom of the bucket. This leaves you with a collection of paraffin-covered insects, and often a lot of lightweight plant fragments, but a quick wash with detergent gets rid of the parrafin. Indeed, I use a standard household detergent that truly does wash twice as many insects as other brands. Recently, the efficiency of this process has been tested by Rousseau (2011) and has been found to recover over 80% of all insect fragments present.

Which orders of insects are preserved and how much they are studied can be variable. Beetles (Coleoptera) preserve well due to their very hard exoskeleton and are usually the most numerous order found in many deposits. They are also the order that, arguably, produces the most useful information due to their frequently restricted ecology (Buckland and Coope 1991; Elias 2010). Admittedly, most of us working in the field are trained to 'do the beetles' first and tend to regard beetle remains as 'home waters' or 'safe territory'.

However, deposits often will contain a range of other insects, with thinner exoskeletons. Both the adults and pupae of flies (Diptera) often are encountered. Fly remains are very useful in the archaeological record, particularly when they come from settlement sites (e.g. Skidmore 1999; Panagiotakopulu 2004). However, their study is still in its infancy. Lice (Mallophaga) and fleas (Siphonaptera) are also sometimes encountered and are considered useful indicators of hygiene and the presence of domesticated animals (e.g. Kenward and Hall 1995; Kenward and Allison 1994). Bugs (Hemiptera) also are routinely encountered, although their study remains specialised (Kenward 2004). Non-biting midges (chironomids) have been used to examine climate and environmental change (Langdon *et al.* 2004; Ruiz *et al.* 2006; Walker and Cwynar 2006).

Identification is not actually that difficult. It is simply a case of playing three-dimensional 'snap' with an archaeological specimen, matching features on ancient specimens against those from a modern reference collection. Key features that are used are:

- Size and shape of the 'whole' fragment
- Large-scale surface features such as the form and location of ridges and punctures
- The nature of the microscopic ground pattern present on the surface
- The pigmentation pattern that gives the ground colour to the fragment. The 'hue', or bright reflective shine that makes beetles glitter, is not reliable since the same species can reflect a number of eye catching, but unhelpful, colours. Still they do brighten a dark day.

Elias (1994, 2010) discusses the methodology of identification in more detail in his recent books. The problem, of course, is the sheer number of fragments and species concerned. Usually it is the head, thorax and wing cases (elytra) of the beetles that are identifiable. So that is at least three different 'bits' per beetle, and there are, at least, 4500 species of beetle in Britain today. Fortunately, this blizzard of possible fragments actually falls into discrete 'sets' of shapes that usually equate to the family or genus levels. Learn the set of basic shapes and one is most of the way there. Getting things to species level is tricky but possible and again needs a very good memory for shapes and the learning of 'tricks' (the artful dodges that mean you can tell one species from another). I am very lucky. One of the few advantages of being severely dyslexic is that I seem to have been given the gift of a 'photographic' memory for shapes as compensation. Years can pass between seeing a particular fragment or illustration and yet I can remember it. At least, if I cannot remember the name, I know where it is in the insect collection.

Where things can go wrong is that your memory can let you down and mistakes can become persistent. For example there are two very closely related species of 'powder post beetle', *Lyctus brunneus* and *Lyctus linearis* both can be separated on the arrangement of punctures on the wing cases. Unfortunately, in my head I switched around which species had the punctures in rows and which had them scattered at random. Not a major mistake perhaps, since both species have very similar ecological behaviour. However, *Lyctus brunneus* is American in origin and has only been established in the U.K. since the early 19th century. The problem is that for several years, in print, I have been suggesting that it may have been present since the Roman invasion. Obviously, this is now rather embarrassing, so if you find a reference to *L. brunneus* in a previous publication of mine please mentally replace this species with *L. linearis*. We will never discuss this again and shall move swiftly onward.

Usually when the taxa are identified the minimum numbers of individuals present are counted up per sample and these are displayed in a very large table. In the case of this book, a summary table like this is included in Appendix 1. This shows a complete taxa list for all of the urban sites from London. The nomenclature used in this table and this book is that of Lucht (1987). I have given the total number of individuals found in all contexts at the sites in column 5. This is followed by the number of individuals in each of the main archaeological periods represented at each site.

In the Chapters that follow, I will indicate just how useful these insects can be when addressing a number of archaeological problems. This is really the aim of this book. There is, however, an ulterior motive. I feel that insect analysis as a discipline has failed to attract large numbers of practitioners. In the UK there are around ten individuals who practice this arcane art with perhaps another dozen working elsewhere in the world. Given that some of these practitioners concentrate only on pre-Holocene deposits, or have recently retired, you can see that we are a rather small club. One very bad bout of food poisoning at an Association for Environmental Archaeology dinner and that is the end of archaeoentomology. This does seem to me to be too small a number to be viable for the maintenance of the discipline in the long term. It also prevents us entering the 'main steam' and becoming a standard archaeological technique. Why so few? I think it results from the difficulty of learning the discipline to some extent, but mostly I feel that we have failed as a discipline to get our message across. I hope that this book makes amends for this to some extent and presents examples of the type of results archaeoentomological data can provide in a digestible way.

A QUICK GUIDE TO THE ASSUMPTIONS, FLAWS AND HISTORY OF ARCHAEOENTOMOLOGICAL INTERPRETATION

How insect faunas from the archaeological record are interpreted is a subject that deserves discussion and certainly more of an airing than it presently receives. Inevitably it also means discussing the work of Harry Kenward of English Heritage and the University of York. Most of the recent ideas about how we should interpret archaeological insect faunas result from his research.

We can roughly trace the evolution of the main types of interpretation used by archaeoentomologists in a chronological order. Of course it is not quite this simple; often ideas have overlapped, or been renamed or clarified by different people at different times.

Archaeoentomology, as a discipline, has its origins with Russell Coope who started working at the University of Birmingham in the 1950s. Coope made the profound observation that if there had been no evolution of insect species in the Quaternary, then the modern ecology of a species could be used to reconstruct past environments (Buckland and Coope 1991; Coope 1978). Of course the clearest example of this concept is the use of environmentally sensitive species to establish past climate (this is discussed further in Chapter 2). This could be labelled as the use of 'indicator species'. Various examples of the use of this technique can be seen in Chapter 4 where the nature of woodland is discussed and for the granary pests in Chapter 5. The pitfalls of an over-reliance on this basic principle are also discussed in Chapter 6.

A further refinement of the 'indicator species' concept is the so-called 'mosaic approach'. This is the logical extension of the use of indicator species and is still viable on many Pleistocene and Holocene sites (Buckland and Coope 1991). Kenward sums up its use thus:

> *...a series of species has been recorded; one discovers what has been written about the habitats of each of them and tabulates the data. A summary of all species' requirements should produce an outline of the ecological conditions near the deposit as it formed with the relative amount of each habitat reflected by the abundance of the species derived from it* (Kenward 1978: 3).

Though I do not think archaeoentomologists deliberately presumed that this was the case, or indeed set out to interpret insect faunas in this simplistic way, but it can be seen to be the 'ghost in the machine' in some early urban archaeoentomology reports. Classic examples are Osborne's (1969) landscape reconstruction at Wilsford Shaft and the early speculative and over optimistic paper by Buckland, Grieg and Kenward (1974) on the first material examined from Anglo-Scandinavian York.

The difficulties of transferring this simplistic approach to the urban archaeological record were clearly outlined by Kenward in his 'Pitfalls to interpretation' paper in 1975b and his *A new approach* volume in 1978, where he identified three problems:

- Often the recorded modern ecology of species is under-researched, contradictory or not relevant to the situation at hand. This is particularly true of many of the insects we find in urban archaeological samples. In terms of urban archaeology this difficulty is perhaps being rectified as we begin to move away from a reliance on modern ecology towards understanding what species were doing in archaeological settlements. More frequently our understanding of the past ecology of a species is being derived from the archaeological record itself. Examples of this latter process can be seen in Chapter 10 in this book.

- Species that have very restrictive ranges of ecology (stenotopes) can be overemphasised in terms of reconstruction. Classic examples of this are discussed in Chapter 6, when the use of 'indicator species' is explored. In particular, species that are quite general in terms of what they do and/or the ranges of habitats in which they live (eurytopes) tend to be downplayed.

- The implications of a series of experiments investigating modern 'death assemblages' (Kenward 1975b; 1978) are perhaps most worrying for the discipline. These beetle faunas were thought to have formed under circumstances similar to those expected for the archaeological record. Kenward found that often there was no actual balance between the real proportions of habitat types in the actual surrounding landscape and that reconstructed from the insect assemblage. The most famous example is from a drain deposit in an alleyway located behind the former buildings of the Environmental Archaeology Unit (University of York) in the middle of York. The insect fauna included species that suggested woodland, decaying plant material, and cattle were present in the local area. This was most definitely not the case.

Kenward (1975b; 1978) blamed these occurrences on the accidental inclusion of 'background fauna' into archaeological deposits. These are species that have been blown, washed or dumped into deposits. Beetles are quite capable of flying over long distances, or walking shorter ones, and getting into deposits that do not represent their normal ecology. After all, we are looking at 'death assemblages' and most things die, not of old age, but because they are in the wrong place. For example the dead dung beetle on your windowsill (and there probably is one, believe me) is not there because you keep cows in your spare room, but because it has flown in (or blown in) from elsewhere whilst looking for a cowpat on which to feed or lay eggs. Like moths, many beetles also are attracted to light and enter brightly lit houses during darkness if given the opportunity.

For a while this seemed a body blow to urban archaeoentomology with the general message of the usefulness of archaeoentomological data somewhat lost due to a frank discussion of the limitations of data. I can remember when I first started to work in the discipline archaeologists said things to me like: *"Why do insects from sites such as this? Kenward has shown that they do not give an accurate* (read truthful*) picture of what was really happening".* In 2009 I had a student in an essay write *'environmental archaeology never recovered from the 1975 Kenward event'.* This is all slightly annoying because:

1) It **is not** what Kenward actually said;

2) Given how often archaeologists will tell us that there is no 'archaeological truth, only interpretation' it rather looked like the pot calling the kettle black.

Kenward has spent the last 20 years refining a number of techniques to try to overcome this difficulty. Initially he devised a number of statistical techniques, such as indexes of diversity and rank order curves (Kenward 1978) to address this problem. The point of these statistics seems to defeat many students, and indeed it took me 10 years to work it out myself. All they really aim to do is to help you distinguish between faunas with a lot of 'background' species and those without this 'background noise'. They help you identify those taxa that are *probably* easier and safer to use in interpretation over others.

Kenward also urged us to begin to devise other ways of looking at the data. The first attempt at this (Kenward 1978) was purely descriptive and consisted of proposing a series of broad 'ecological groups' which were directly useful to the archaeological record. Membership of the early groups was mainly determined by what was known from a species modern ecology. This culminated in the devising of a set of ecological codes that were first used systematically with the analysis of the Roman deposits from the *Colonia* at York (Hall and Kenward 1990). This set of groups is outlined in Table 1.1 and the individual memberships of these groups for the studies from London are indicated in column 3 in Appendix 1. Kenward's groups included a set that appeared to outline differing degrees of breakdown in organic matter. It also suggested that there was a group of species, the infamous 'house fauna', that seemed to always be associated with settlement in the past. This was the genesis of trying to use actual archaeological association of insects with each other, rather than their modern ecology, as the primary tool of interpretation. By the time the Coppergate volume (Kenward and Hall 1995) was published two further groupings based on the analysis of the archaeological record were proposed. This was a 'subterranean/ post-depositional' community and the 'oxyteline association'. Subsequently, Kenward has used a number of statistical techniques to define and refine these groupings using data from the excavations at York (Carrott and Kenward 2001). This will be discussed in more detail in Chapter 10 but these 'new' groupings are also outlined in Figure 1.1. Similar attempts to use advanced statistical techniques to define meaningful patterns for use in interpretation can also be seen in work by Perry *et al.* (1985) in Iceland.

What is clever about this technique is that it uses the archaeological record and the insect faunas to produce patterns of interpretation. To some extent this mitigates for the difficulties of using the modern ecology of species, the presence of background fauna and the fact that many deposits in the archaeological record no longer have modern equivalents. A short hand is to refer to groups established in this way as 'indicator groups'. These will be used widely throughout this book, but probably to best effect in Chapter 8.

This approach naturally leads us to think about the sensible idea of combining these 'indicator groups' with the results of other forms of archaeological analysis. Usually this includes plant macrofossil and pollen analysis along with the archaeological record itself. This appreciation of the need to work in tandem with other specialists can be seen from the earliest work of Coope (Buckland and Coope 1991), and was used in Buckland *et al.* (1974) but has only recently been formalised in papers by Kenward and Hall (Kenward and Hall 1997; Hall and Kenward 2003, Hall and Kenward 2011) where they suggest a number of 'indicator packages' which provide interpretation based on a whole range of biological and archaeological evidence rather than relying on insects alone. This will be discussed further in Chapter 10.

In this book I have used groupings based on Hall and Kenward (1990) but mainly derived from those included in the 1995 Coppergate volume (Kenward and Hall 1995). It is also worth noting that Mark Robinson has developed a set of groupings that can also be used to help reconstruct landscapes as well as settlement deposits (Robinson 1981; 1983)

SOME THOUGHTS ABOUT URBAN ARCHAEOENTOMOLOGY

Lewis Binford (1981) labelled one of the commonest traps in thinking about the archaeological record as falling into the 'Pompeii premise'. This assumes that what we observe in the archaeological record is the last action carried out at the site. This is the case at Pompeii where the events of the last few hours of the Roman town are preserved as the result of a cataclysmic event, the eruption of the volcano which effectively sealed the site and stopped any later alteration of the archaeological record by humans. This situation, he suggests, is in fact very rare archaeologically. Most archaeological materials, and archaeological sites, have a life after their main phase of use. Materials and deposits can be removed and reused. Decay and change of the site after abandonment also can completely alter its nature. For an example, go and visit Kenilworth Castle in the Midlands. What you see is not how the building looked in the 15th century. It has been partially demolished, material has been 'robbed' or reused. Various parts of the ground surface have been levelled and paved so that it can be used for its present purpose (tourism).

This is a problem we often face in urban archaeoentomology and we need to think more about it. A classic example of the trap of the 'Pompeii premise' is how we think about the insect evidence for 'living floors' in buildings. There is a tendency to see the insect faunas from these deposits in the literature as indicating that the flooring during occupation was very squalid and deeply unpleasant. That the build up of rotting floor material, organic matter and waste may have been so deep that 'you lost the children in it'. This culminated in an idea suggested by Russell Coope (1981) that this might have been deliberate and the heat generated by the decay may have acted as a form of 'central heating'.

This seems unlikely to me. I think it fails to consider what happens to archaeological sites and archaeological deposits after their primary use. We need to think about the following issues:

Abandonment

Strange things seem to happen to archaeological buildings after they are abandoned. I suspect that they become the dumping ground for rubbish from a variety of sources. After all, why take stuff a long distance away when it can be 'tucked' into a nearby conveniently abandoned house? Several years ago, Siobhan Geraghty showed a series of slides at a conference that beautifully illustrated this point. Some builders left a skip in the courtyard at the Museum in Dublin. First building waste appeared, and then the broken computers from peoples' offices, then the gardening waste from home and so on. I even remember the traditional supermarket trolley in the photos, but I might be embellishing. The point she was making is that a lot of 'stuff' could accumulate after any building falls out of use that has no direct reflection on its primary use. Certainly Lisa Moffett and I felt this was the origin of the 'floor deposits' we examined from Stone Staffordshire (Moffett and Smith 1996). However, there are clear cases where this is not an issue. In particular, many of the house floors from medieval settlements in Greenland and Iceland appear to have been abandoned so suddenly and completely, that the build-up of detritus after abandonment could not be a factor (Buckland *et al.* 1983; Buckland *et al.* 1992).

Re-deposition

This is a common problem recognised widely in archaeology (e.g. Rowley-Conwy 1984). In studies of pottery it is referred to as 'residuality'. In palaeoenvironmental and geoarchaeological studies this situation often is called 're-working'; when materials or objects are removed from the location in which they were first used or formed and become included into younger deposits. The problem is that the re-deposited material can have no similarity in terms of use, formation or content to the new deposit into which they ultimately are incorporated. Materials can also be re-deposited by human action as archaeological sites are formed. In particular, the constant digging up, and indeed churning up, of an archaeological site, as buildings are repaired/ redeveloped or pits are dug, seems particularly relevant. Over a short period of time, several differing contexts and materials can be 'blended' together. This issue will be discussed further in Chapter 7. However, on some sites, notably those at York (Hall and Kenward 1990; Kenward and Hall 1995), this does not seem to be a factor with many house floors appearing to be relatively intact.

Re-deposition can occur as the result of natural processes as well. This is less of a problem for settlement deposits but is crucial for deposits such as sands, silts and 'peat' in abandoned river channels (Howard and Macklin 1999). Here river bank-side erosion can result in material several thousand years old becoming incorporated within much younger deposits. This is a real headache when attempting to date or carry out environmental analysis in this type of active alluvial system.

Decomposition after burial

This is an issue that has been raised repeatedly by Harry Kenward (Hall and Kenward 1990; Kenward and Hall 1995; Carrott and Kenward 2001). He wonders if many of the insects we see in the archaeological record were not living in the material when it was initially formed but invade it after it was buried. This could be the factor that explains why so many urban deposits appear to be so nasty in terms of the nature of the insect faunas recovered. The 'yuck' factor results not from the use of the material by humans but simply from what happened next. Kenward (Kenward and Hall 1995, Carrott and Kenward 2001) suggests that there is a distinct group of species that occupy buried material. This is the 'post-deposition' insect fauna suggested by Kenward (Hall and Kenward 1995) and, to some extent, equivalent to 'Group S' in the scheme of Carrott and Kenward (2001).

Are the faunas representative?

This is a major concern in any discussion of the archaeological record. Just how typical are the living conditions suggested by the insect faunas for settlements as a whole? One way of thinking about this is to remember that the sites on which we work have to be waterlogged. Often the waterlogging occurs because the site is near to the edge of a river or it is in a low-lying area of town. This is clearly the case with many of the sites in London examined in this book. The use of wet, low-lying areas may not be typical of the settlement as a whole. Are the squalid conditions suggested by the insect faunas representative of 'rising damp' and surface water at these particular locations rather than being typical of conditions in most London housing? Are the activities seen in these areas, such as dying and tanning, typical of the activity in most parts of town? What socio-economic 'strata' of people lived and worked in the wetter areas of town? How typical are they and the conditions they lived in of the community as a whole? Think about this. If in 2000 years time the only parts of London to be excavated are the worst part of the modern 'sink estates', what kind of reconstruction might result? Certainly different from the impression gained should we only excavate areas of Knightsbridge.

However, a raised watertable is not always the cause of waterlogged preservation in urban deposits. For example, at Anglo-Scandinavian York it is thought that the preservation occurs not as the result of a wet ground surface but rather the deposition of large amounts of organic matter causing a 'sponge effect' which pulls in water (Kenward and Hall 2006).

Now, before we go further, I want to make it plain that I am not setting out to darken the name of urban archaeoentomology, or the role that it can have in archaeological interpretation. These are problems that are common to all palaeoenvironmental and archaeological disciplines (Lowe and Walker 1997a, Bell and Walker 2005). It would be a poor discipline that did not own up to its own problems or discuss them. It is the acknowledgement and overcoming of these difficulties that is the challenge. It is also the thing that makes the archaeological games we play fun. One obvious solution to this conundrum is to adopt a multi-proxy approach (Kenward and Hall 1997) to the sampling of deposits such that archaeological, and indeed environmental, interpretation/ reconstruction can be drawn from multiple lines of independent evidence.

Figure 1.1. Common schemes of grouping the ecology of insects from archaeological sites with the nearest ecological equivalents between them (Black areas are not discussed in the scheme)

Coding and interpretation	Common members	Coding and interpretation	Common members	Coding and interpretation	Common members
oa (& ob) species which will not breed in human housing	e.g.: Carabidae, Scarabidae, Curculionidae	Group from Hall and Kenward 1990 repeated			
oa – a aquatic species	e.g.: Dyticidae, Hydreanidae, Hydrophilidae, Drypoidae	Group from Hall and Kenward 1990 repeated			
oa – c species associated with salt water and coastal areas	e.g.: *Tachys scutellaris, Cercyon depressus*	Group from Hall and Kenward 1990 repeated			
oa – ws species associated with damp watersides and river banks	e.g. *Agonum spp., Cercyon convexiusculus, Olophrum spp., Lesteva spp,* some *Trogophloeus spp. Some Platystethus spp,* some *Phalacrus, Donacia, Plateumaris, Tanysphyrus lemnae, Notaris spp, Limnobaris spp.*	Group from Hall and Kenward 1990 repeated		Group B Species commonly associated with gullies and wet external areas	*Cercyon analis, Omalium rivulare, Oxytelus rugosus, Trogophloeus bilineatus, C. fuliginosus, Oxytelus sculptus, Platystethus arenarius, Neobisnus* spp, *Lithocharis* spp., Some *Monotoma* spp., *Anthicus* spp., (overlaps and has other member in group C and D)
rd species primarily associated with drier organic matter.	e.g. Lathridiidae, Cryptophagidae, *Typhaea stercorea, Aglenus brunneus, Mycetaea hirta,* Ptinidae	Group from Hall and Kenward 1990 repeated		Now part of Group A	
rf Species primarily associated with foul organic matter often dung.	*e.g. Cercyon impressus , C. haemorrhoidalis , C. melanocephalus, C. unipunctatus, C. atricapillus* (Marsh.), many Histeridae, *Platystethus arenarius, Anthicus* spp., *Geotrupes* spp, *Onthophagus* spp, *Aphodius* spp.	Group from Hall and Kenward 1990 repeated		Group C Species probably associated with external organic rich 'muds' and quite foul deposits like in- filled gullies and the edges of pits	*Cercyon analis, Omalium rivulare, Oxytelus rugosus, Oxytelus complanatus, A. nitidulus, Platystethus arenarius, P. cornutus* (group), *P. nitens, Gyrohypnus* spp., *Oxyomus silvestris,* Some *Aphodius* species (overlaps and has other members in group B and D)
rt Insects associated with decaying organic matter but not belonging to either the rd or rf groups.	*e.g. Cercyon analis,* many *Staphylinidae* including *Xylodromus concinnus, Omalium* spp., *Oxytelus* spp. *Rhizophagus., Monotoma* spp., *Trox scaber*	Group from Hall and Kenward 1990 repeated		Group D Species probably associated with very foul deposits in pits	*Cercyon unipunctatus, C. analis, C. haemorrhoidalis, Omalium rivulare, Platystethus arenarius.*
		Oxyteline association – probably associated with wet, muddy areas filled with organic detritus	*Trogophloeus bilineatus, T. pusillus; T. fuliginosus, T. corticinus, Oxytelus complanatus, Oxytelus rugosus, A. nitidulus, A. sculptus, Neobisnus*	*Now part of groups B and C*	

				Group E Cesspits	*Patrobus atrorufus, Coprophilus striatulus, Tachyporus* spp., *Cercyon haemorrhoidalis, Ptenidium* spp., *Clambus* spp. *Omalium rivulare, Oxytelus teracarinatus, Corticaria* spp., *Mycetaea hirta, Gracilia minuta, Cidnorhinus quadrimaculatus*
				Group F Probably associated with wool processing and sheep 'dags'	*Aphodius granarius, A. Prodromus, Geotrupes* spp., *Damalina ovis, Pulex irritans*
		Subterranean /post depositional	*Trechus micros, Coprophilus striatulus, Trichonyx sulcicollis Rhizophagus parallelocollis*	Group S Subterranean decomposer community	*Trechus micros, Coprophilus striatulus, Rhizophagus parallelocollis, Trichonyx sulcicollis*
g species associated with grain	*e.g. Oryzaephilus surinamensis* , *Laemophloeus ferrugineus, Sitophilus granarius*	Group from Hall and Kenward 1990 repeated			
l species associated with timber	*Lyctus linearis, Anobium punctatum, Xestobium rufovillosum,* Scolytidae.	Group from Hall and Kenward 1990 repeated		Now part of Group A	
P phytophage (plant feeding) species often associated with waste areas or grassland and pasture	A range of Elateridae, Chyrsomelidae and Curculionidae	Group from Hall and Kenward 1990 repeated		Some included in Group A	
h members of the 'house fauna': this is a very arbitrary group based on archaeological associations (Hall and Kenward 1990).	*Xylodromus concinnus,* Lathridiidae, Cryptophagidae, Typhaea stercorea, *Aglenus brunneus, Mycetaea hirta, Ptinidae Lyctus linearis, Anobium punctatum*	Group from Hall and Kenward 1990 repeated		Group A Settlement deposits	*Xylodromus concinnus, Coprophilus striatulus,* Lathridiidae, Cryptophagidae, *Typhaea stercorea, Aglenus brunneus, Mycetaea hirta, Ptinidae Lyctus linearis, Anobium punctatum, Pulex irritans, Blaps, Tenebrio obscurus, Damalina ovis, Melophagus ovinus, Trox scaber, Sitona spp.,* (and other members of groups C, E and F)
Pu species associated with pea, beans and other pulses	*Bruchus* spp.	Group from Hall and Kenward 1990 repeated		Now part of Group E	

CHAPTER 2: LONDON BEFORE THE CITY: THE ICE AGES

INTRODUCTION

This chapter and the next provide an overview of how the landscape of the Lower Thames Valley has changed between *ca.* 350,000 years ago and the arrival of the Romans in *ca.* 50 AD; which roughly equates to the periods known as the Pleistocene and Holocene (from *ca.* 10,000 years ago) in England. This chapter examines the role that insect analysis has had in reconstructing London's Pleistocene past and also outlines how archaeoentomologists work on deposits of this age. It also introduces some of the major issues and problems facing archaeoentomologists in terms of reconstructing ancient landscapes. Chapters 3–9 specifically address the last 10,000 years, which corresponds to the period known as the Holocene (or the Flandrian).

WHY DO WE HAVE SO FEW INSECT FAUNAS IN GREATER LONDON BEFORE THE ROMANS?

What makes for a good 'landscape archaeology' insect project? Archaeological landscapes where a considerable range of sites and samples have been analysed have allowed a detailed, continuous and coherent picture of how those landscapes develop and change through time. Examples include Thorne Moor and Hatfield wastes in South Yorkshire (Buckland 1979; Whitehouse 1997, 2000, 2004), the Trent Valley on the Derbyshire and Leicestershire boarder (Knight and Howard 2005; Greenwood & Smith 2005), the Somerset Levels (Girling 1977; 1979a; 1980; 1982; 1985) and the Gwent Levels (Smith *et al.* 2000; Tetlow 2004). A factor that has made these studies good examples of landscape archaeology and palaeoenvironmental work is that the insect analysis is only one component of a wider study. These studies include a range of other palaeoenvironmental techniques; such as pollen and plant macrofossil analysis, and a sensitive use of science-based dating techniques including radiocarbon determinations and dendrochronology, in addition to archaeoentomological analysis. This, of course, all works to much greater effect if the area in question produces a wealth of artefact, settlement and monumental archaeology. This gives the insect faunas a 'human dimension' and academic research validity.

Unfortunately, none of these conditions really apply to the Pleistocene insect faunas from Greater London. Despite the size of the area concerned (see Figure 2.1) the number of sites where insect remains have been analysed is actually extremely limited. In addition, the sites are spread out thinly across a considerable period of time.

Furthermore, most of the work undertaken does not come from the area of the 'city' itself but from sites to the west and east of London. Why is there such a shortage of Pleistocene sites? There would seem to be one obvious explanation for this paucity: the presence of the city itself. Large-scale strategic archaeological investigation is difficult within the city due to both the physical barrier of the 'built environment' of modern London and the presence of a deep 'layer cake' of post-Roman urban archaeology beneath existing buildings. London's archaeology also has to wait for development to occur or rely on chance finds. All we have been able to gain, therefore, are small and dispersed 'snap shots' of the Pleistocene (450,000 – 10,000 years ago) and Holocene (the last 10,000 years) landscapes of London. Merriman (2000, 35) holds that, as a result, the prehistory of the city

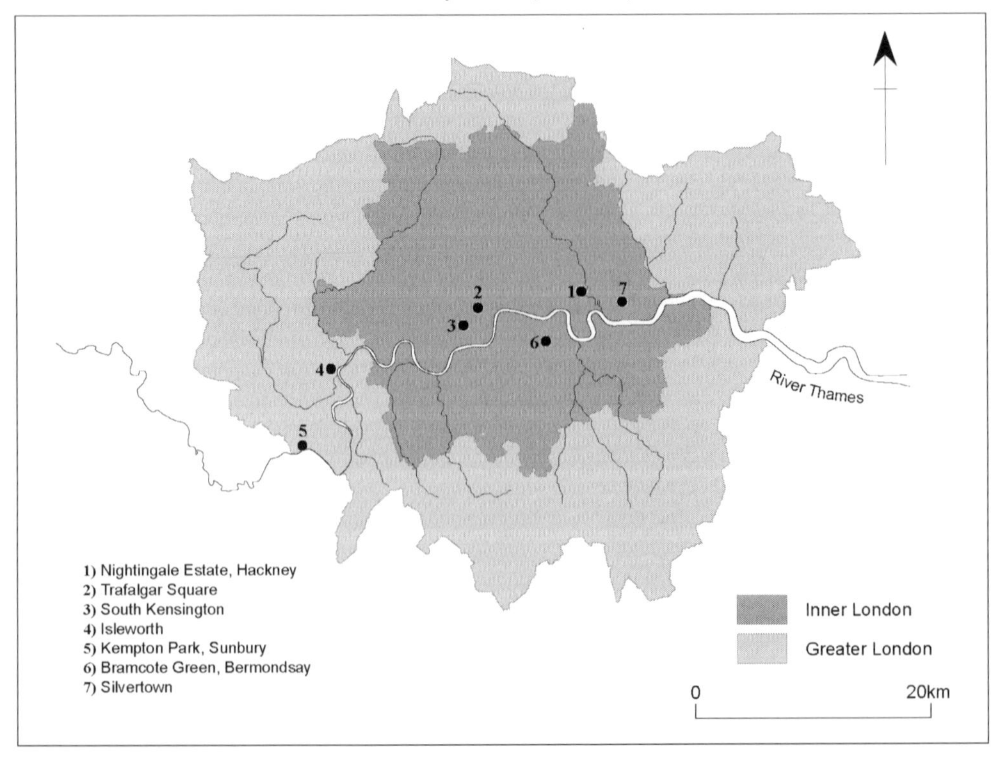

Figure 2.1. Location of the Pleistocene sites in Greater London mentioned in text

itself is 'a meaningless entity in prehistoric terms'. This might lead to the view that there is a wealth of hidden prehistoric archaeology and Pleistocene geology to be found and sampled for insect faunas if only we could remove the modern city. There is, however, another way of looking at this: we could accept that London was a bit of a backwater in prehistoric times. Until the last few years one striking aspect of the prehistory of Greater London is actually how little there is compared to other areas of the British Isles (Merrifield 1975; Merriman 1990, 2000; Cotton 2000). However, in the last decade a number of both Neolithic and Bronze Age sites have been excavated, even within the city boundaries, suggesting that the area may have been more populated in the past than once thought (Jane Sidell *pers. com.*)

BACKGROUND: LANDSCAPE AND GEOLOGY

The London basin is a wedge shaped depression bounded by the Chilterns to the north, the North Downs to the south and the Berkshire downs to the west. Chalk and the 'London clay', capped in places by the 'Bagshot' Sands, underlie the depression (Rackham and Sidell 2000). On top of this 'hard geology' is a collection of 'superficial' deposits laid down during the Anglian glacial period (Marine Isotope Stage 12) *ca.* 480,000 – 430,000 years ago (that's superficial in the geological sense but not in terms of the story we wish to tell here). To the north these deposits include areas of till that marks the southernmost extent of glaciation and in central London, and to the south, this period is represented by a series of gravel spreads. These gravels were initially produced as outwash from the glaciers and subsequently cut into a series of low terraces by the River Thames. Naming, dating and categorising these terraces have become a good form of sport for the geologists of London over the years culminating in Bridgeland's (1994) scheme. Scattered throughout these deposits are lenses of organic clays and silts that developed in low swampy depressions, or as the remains of abandoned river channels. These are typically the deposits that produce the remains of insects, plant material and pollen that are important to the story of the early prehistory of London. The dating and naming of the interglacial periods (warmer periods between cold ice ages) that these deposits come from has also become something of a field sport. This is particularly true since there are disagreements between the British and the European sequence (Lowe and Walker 1997a). However, the chronology has become easier to understand since these events are now normally tied to the various oxygen isotope fluctuations recorded from sediments in deep ocean cores (Lowe and Walker 1997a). The major highs and lows of this scheme are outlined in Figure 2.2.

USING INSECTS TO RECONSTRUCT LANDSCAPE, CLIMATE AND TO DATE DEPOSITS: AN EXAMPLE USING THE OLDEST INSECT FAUNA FROM LONDON

Serendipitously, amongst the small number of faunas from London are some of the oldest in the country and some of the most important to the founding of 'modern palaeoentomology'. The Trafalgar Square deposit, mentioned below, was amongst the first examined by

OI STAGE	EPOCH	STAGE	PERIOD	FLANDRIAN CHRONOZONES	GODWIN ZONES	CULTURAL PERIODS	CALENDAR YEARS BC/AD
One	Holocene	Flandrian	sub-Atlantic	Fl III	VIIc	Post-medieval	
						medieval	AD 1000
						Saxon & Danish	
						Roman	0
						Iron Age	
			sub-Boreal		VIIb	Bronze Age	1000 BC
						Neolithic	2000
							3000
							4000
			Atlantic	Fl II	VIIa	Mesolithic	5000
							6000
			Boreal	Fl Ic	VIc		
				?	VIb		7000
				Fl Ib	VIa		8000
					V		
			pre-Boreal	Fl Ia	IV		9000
Two	Pleistocene	Devensian	Loch Lomond stadial (Younger Dryas)		III	Upper Palaeolithic	10,000
			Windermere interstadial (Allerød)		II		11,000
			Dimlington stadial (Older Dryas)		I		12,000

Figure 2.2. Table of major period and dates for the Pleistocene (reprinted with permission from Sidell *et al.* 2000)

Prof. Russell Coope (then based at the Department of Geology, University of Birmingham). It was also during the early work on sites such as this, along with those at Upton Warren and Chelford (Coope 1959, 1961), that it became clear that the insects found in Pleistocene dated deposits were identifiable and that no detectable 'speciation' had occurred during this period (Buckland and Coope 1991). Nearly all the taxa of beetles so far found in Pleistocene deposits exist today (Buckland and Coope 1991). You may have to hunt though the more obscure drawers of the world's insect collections for modern comparatives but they are there. Perhaps the most dramatic example of this is the small dung beetle that kept occurring in glacial deposits from the last Ice Age. For many years the identity of this species remained a mystery until a casual glance by Russell Coope into the drawers of the Natural History Museum's collections revealed its identity. It was *Aphodius holderi*, which is found today in a limited number of high valleys in Tibet, but whose fossils are present in a wide range of ice age sites in Lowland Britain (Coope 1973).

Coope (1959, 1961) also quickly made the interpretive step that linked the modern ecology of the species to their past behaviour, allowing reconstruction of a detailed picture of the past climate, environment and landscape. In addition, because of consistent results, he argued that the assemblages of fossil beetles can sometimes be used to roughly date deposits (Coope 1959, 1961, 2001).

In this section I will show how this is achieved using the oldest insect fauna from Greater London, the Nightingale Estate fauna from Hackney (Green *et al.* 2004). This site has been dated to Marine Isotope Stage (MIS) 9 - the Purfleet Interglacial – *ca.* 350,000 years ago. Nationally it is only surpassed in age by the faunas from Pakefield, Norfolk (Parfitt *et al.* 2005; Coope 2006) that date to MIS 17 – the early part of the Cromerian Interglacial – 700,000 years ago and a series of sites dating to the MIS 13 – the Late Cromerian – 500,000 years ago (Coope 2006). Slightly nearer in time are two Midlands sites at Nechells, Birmingham (Shotton and Osborne 1965) and Stretton on Dunsmore, Warwickshire (Keen *et al.* 1997) which date to MIS 11 - the Hoxnian Interglacial – *ca.* 400,000 years ago.

At Nightingale Estate the insects provide an extremely vivid reconstruction of the landscape present in this part of the London Basin 350,000 years ago (Green *et al.* 2004). It appears that a large and rapidly flowing gravel-bottomed river ran through an area of dense woodland (Green *et al.* 2004). The forest contained oaks (*Quercus*), hazels (*Corylus*), and other deciduous trees. Many of these trees were dead and dying and deadwood littered the forest floor. Some of the trees may have been thick with the growth of old man's beard (*Clematis vitalba*), a clinging vine. There were also meadow-like glades full of vetches and clovers that were grazed by large herbivores. Other beetles suggest that the patchwork of habitats present also included areas of sandy heath covered by heather. Near to the river there was carr, or swampy woodland, which included alder (*Alnus*), willow (*Salix*) and birch (*Betula*). Reed sweet grass (*Glyceria*) grew along the riverside with stiller patches of the river covered in duckweed (*Lemna*), water milfoils (*Myriophyllum*) and water-lilies (*Nymphaea*). This pleasant sounding landscape also had an equitable climate. Summer temperatures were considerably warmer than today with mild winters no more severe that we suffer at the moment. Rainfall was probably similar to today, with river levels high year round.

How do we know the temperature at the time?

Climatic reconstruction, particularly of yearly temperature ranges, is perhaps one of the most important outcomes of the analysis of insect remains from Pleistocene sites (Elias 2010). The theory and process that underlies this analysis has been extensively explored in Elias (2010). Such reconstructions rely on one very strong assumption: if you know the modern temperature range of an insect species this suggests that the same conditions must have existed at the sampling location in the past. These 'temperature range envelopes' are limited only by our knowledge of the modern distribution of the species. A key example of this is *Boreaphilus henningianus* from the Late Glacial deposits from the Glanllynnau kettle hole (Coope and Brophy 1972). Its modern distribution is shown in Figure 2.3 and it is essentially limited to the mountains of Northern Scandinavia and Siberia.

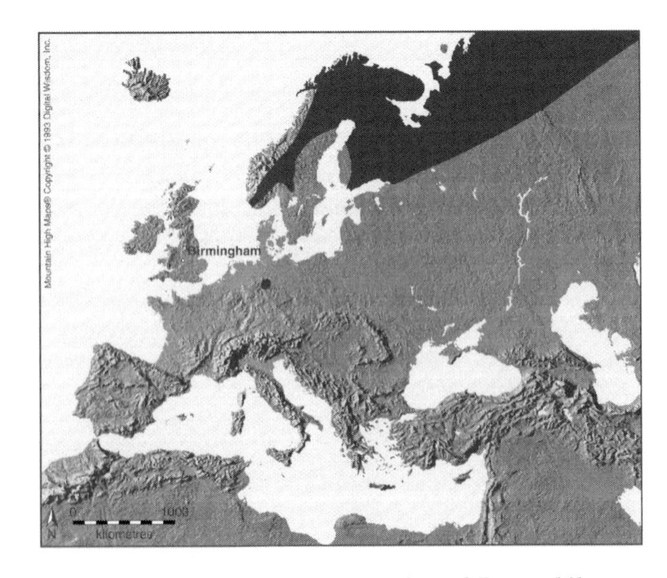

Figure 2.3. The modern distribution of *Boreaphilus henningianus* (redrawn from Coope and Brophy 1972 and Gaffney *et al.* 2009)

The distribution of beetles appears to be controlled by three factors: the maximum summer temperature (T-MAX), the minimum winter temperature (T-MIN) and the range between these two temperatures (T-Range). In 1959 Russell Coope produced the first temperature estimation for the late glacial from the Chelford site (Coope 1959) and this work eventually lead to the initial version of his temperature curve for the last glaciation, which was published in 1977 (Coope 1977). At the time these results were greeted with a great deal of scepticism but have come to be widely accepted and are now backed up by temperature reconstructions based on ice core and deep ocean sediment analysis (Lowe and Walker 1997b). A further refinement of this technique came in 1987, when along with co-workers Coope used the 'mutual climatic range (MCR)' of a number of species at each site to increase the 'resolution' of the climatic reconstructions produced (Atkinson *et al.* 1987). The use of the MCR has produced an elegant and continuous temperature curve for the last 16,000 years (Figure 2.4). There are also discontinuous, isolated temperature estimations for other periods, as far back as 400,000 years in the British Isles. Recently there have been attempts to increase the 'resolution' and security of the MCR analysis (Bray *et al.* 2006; Elias 2010).

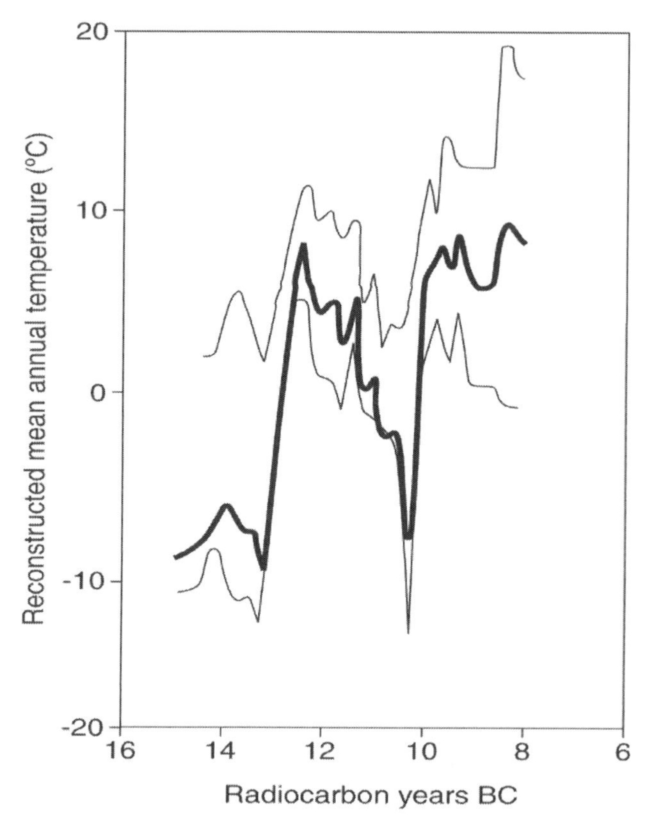

Figure 2.4. Temperature curve for the British Isles established using insect faunas (redrawn from Atkinson 1987 and Gaffney *et al.* 2009)

Today, climate reconstruction based on Arctic ice cores and from deep sea sediments have captured the popular and academic imagination. However, they can only tell us about general global tends and allow an extrapolation of worldwide temperature thresholds. They cannot tell us the temperature at a particular spot in the Midlands of England 22,000 years ago. The only way of gaining accurate temperature estimates for 'inland' areas is by using insect remains; other fossil insects such as chironomid 'midges' also may have a role to play here.

So to return to the Nightingale Estate case study, Figure 2.5 demonstrates how in practice climate can be reconstructed through the mutual climatic range of beetles present (Green *et al.* 2004).

How do we know the details of the landscape?

Landscape reconstruction is mainly achieved by looking at the modern ecology of the individual species of insect identified. These data often can be rather 'general', for example, suggesting broadly that 'pasture and grassland' or 'sandy ground' is present. Sometimes it can be much more specific. For example, the plant feeding insects (phytophages), such as the weevils, are frequently restricted to only one, or a few, species of 'host plant'. Often these plant species only occur in a certain environment or habitat. Figure 2.5 outlines how the insect fauna recovered indicated the various elements of the reconstructed landscape of Nightingale Estate.

In terms of technique and analysis the common practice is to build up the 'picture' by starting with the most abundant insects present, presuming these represent the most dominant types of landscape. Individual species with a specific story to tell, often called 'indicator species', are then referred to in order to support the general picture when relevant. This is an extension of the 'mosaic approach' defined and criticised by Kenward (1978) that was discussed in Chapter 1. The inherent weakness of this approach is that although there is clear evidence for the presence of each element in the 'mosaic of environments' it actually is very difficult to suggest the relative proportions of each habitat type, where they were, or even which may have been dominant in the overall landscape. This is a point that we will return to later when we discuss the nature of Early Holocene woodlands at the end of this chapter.

How do insects help us date the deposit?

Dating of Ice Age and interglacial deposits can be extremely difficult. Normally 'science-based' dating techniques, which depend on physics or chemistry, are used to produce definitive dates for deposits. However, the dating of the deposits from Nightingale Estate, Hackney demonstrates the difficulties that can be encountered and how insect analysis can help (Green *et al.* 2004). Of course, the most commonly used form of science-based dating technique is radiocarbon determination. Unfortunately, radiocarbon has an upper age limit of *c.* 40–50,000 years (Aitkin 1990) and, therefore, it has little relevance for older deposits. Additionally, there is a large gap in time between the end of the period covered by radiocarbon dating techniques and the start of that covered by a range of other isotopic dating techniques such as Potassium/Argon. This leaves us with a number of science-based dating techniques such as Optical Stimulated Luminescence (OSL) and Amino Acid Racemization (AAR) that produce dates with large age ranges and which often fail to provide consistent dates. In the case of Nightingale Estate the OSL dates suggest MIS 9 and the AAR dates MIS 7. This is a difference of some 200,000 years. To resolve the chronological tangle at Nightingale Estate, two more 'archaeological' techniques were used: stratigraphy and biostratigraphy.

Conclusions reached from the insects present	***Significant insects***	**Ecological data these species provide**
A warmer summer climate than today	45 species in the fauna have 'climate envelope' estimates in the MCR database.	Mean annual temperature is about 18°C (though several of the species below are not in the MCR so summer temperature was probably a couple of degrees higher).
	Bembidion elongatum, Chalaenius sulcicollis, Microlestes minutulus, Pelochares versicolor, Caccobius schreberi, Aphodius carpetanus, Chrysomela limbata, Scolytus carpinii, S. koenigi, Rhyncolus punctulatus, Stenocelis submuricatus, Tridonata spp.	The modern geographical range of these species does not reach Britain and most are limited to southern Europe suggesting that a climate similar to that area today existed.
	Bembidion octomaculatum, Abax parallelus, Cybister lateralimarginalis, Onthophagus taurus, Onthophagus fracticornis, Rysselmus germanus, Pleurophorus caesus	These species do not occur in old collections or have very limited distributions in the British Isles. This suggests warmer temperatures and probably continental conditions.
Precipitation must have been relatively high	The large numbers of 'Elmid' riffle beetles.	These species take the whole year to complete their life cycle and need permanent water all year round
The river is large, fast flowing and has a gravel bottom	*Oulimnius troglodytes, Normandia nitens, Limnius volckmari, Oulimnius tuberculatus, Esolus parallelepipedus, Elmis aenea, Riolus cupreus, Stenelmis canaliculatus, Helichus substriatus*	All of these species of 'riffle beetle' are associated with fast flowing and clear water. All prefer a sandy or gravel river bed and do not tolerate silt.
	Hydropsyche contubernalis, Hydrophsyche pelucidula and Cheumatopsyche lepida	The larvae of these species of Caddis fly are normally associated with larger rivers and fast flowing rivulets
The presence of an old mature woodland full of dead wood	*Valgus hempiterus, Cerylon histeroides, Aspidiphirus orbiculatus, Eledona agaricola Pogonocherus hispidulus Anobium spp., Dryophilus spp, Scolytus keonigi, Xyleborus dispar, Dryophthorius corticalis, Rhyncolus chloropus, Rhyncolus puctulatus, Stenoscelis sumuricatus*	Amongst these species are a range of 'bark' and 'shot borer' beetles and weevils that prey on small insects under the bark of dead trees. A number of these species feed on dead wood infected by fungus.
The species of trees that were present in the woodland	*Rhynchaenus quercus* (a 'leaf miner' on oak), *Aploderus coryli* (hazel), *Rhamphus pulicarius* (willow, birch, and alder buckthorn), *Agelastica alni* (alder).	These species of beetle feed on the leaf of a particular species of tree. (The various tree 'hosts' for these species have been indicated after the species name).
	Lucanus cervus (stag beetle; mainly oak), *Scolytus koenigi* (maple), *Platypus cylindrus* (mainly oak)	These species feed on the dead wood of specific trees.
	Curculio venosus (oak) *Culclio nucum* (hazel)	These species are 'nut weevils' which feed on fruits of specific trees.
Old man's beard is present	*Xelocleptus bispina*	This 'shot borer' attacks only the stems of traveller's joy (*Clematis vitalba* L.).
Grassy clearings are present	*Zabrus tenebriodies, Agriotes obscurus, Adelocera murina, Melolontha melolontha, Serica brunnea, Tridonta spp.*	These beetles all feed as larvae on the roots of grass or plants common in meadowland.
	Longitarus, Haltica, Chaetocnema, Sitona and *Bruchus* species	These beetles feed on a range of weedy plants common in grassland.
Clearings were grazed by large herbivores	*Copris luaris, Caccobius schreberi, Onthophagus taurus, Onthophagus fracticornis, Aphodius carpetanus, Aphodius conspurcatus, Heptaulacus sus, Heptaulacus testudinarius*	These are all species of scarab 'dung beetles' which either consume the dung of large animals or feed on the other small insects present. This was quite a dominant feature of the fauna perhaps indicating large numbers of grazing herbivores.
The presence of heath	*Microlestes minutulus, Ryssemus germanus, Pleurophorus caesus.*	These are all species that live in dry sandy areas which are found in heathland
	Micrelus ericae	This weevil feeds only on purple and white heather (*Erica* and *Calluna* spp.) which is common on heathland
The river also contained slow areas surrounded by reeds and covered with floating vegetation?	*Gyrinus* species (whirligig beetle), *Hydrochus elongatus, Coelostoma orbiculare, Enochrus species, Hydrophilus caraboides*	These 'water beetles' are all associated with stationary water suggesting areas of slow water associated with the river channel. These occur in very small numbers suggesting that this was not a dominant feature of the riverside environment.
	Donacia semicuprea	This 'reed beetle' usually feeds on the reed sweet grass (*Glyceria maxima*) that grows in slow water areas.
	Macroplea appendiculata (pondweed and water milfoils), *Donacia crassipes* (water-lilies), *Tanysphyrus lemnae* (duckweed)	These plant feeding species are all associated with plants that float on the surface of slow or still waters (the specific hosts are indicated after the species name).

Figure 2.5. The ecology of the significant insects from Nightingale Estate, Hackney and their interpretation

Stratigraphy is one of the oldest of geological (and archaeological) dating techniques. This refers to the eminently sensible geological principle that the oldest deposits are the lowest; otherwise known as the law of superimposition. In the case of Nightingale Estate some complicated stratigraphic arguments had to be made to suggest that this deposit belonged to the same period as other Hoxnian Interglacial terraces in the area (Green *et al.* 2004). In particular Bridgeland's (1994) scheme of terrace stratigraphy for the London area would have to be adjusted and reinterpreted. One of the strongest reasons why Green *et al.* (2004) felt confident in this radical proposal was that the insects, plants and molluscs present more closely resembled those of other MIS9 sites, rather than the alternatives of MIS7 or MIS 11. Using biological remains to approximately date deposits is called biostratigraphy. Until the development of 'science-based dating techniques' this was the main method of dating geological deposits, although this has subsequently fallen out of use as more precise chronological determination techniques have developed. In the specific case of the insects from Hackney, the assemblage contained a wider and different range of species than that of the later Aveley Interglacial (*c.* 230,000 – 180,000 – MIS 7) (Green *et al.* 2004). Many of the taxa had a pronounced southern distribution and also are very different in their nature to those seen in the preceding Hoxnian Interglacial (c. 430,000 – 380,000 – MIS 11) (Green *et al.* 2004). How do the Nightingale Estate insects compare to other MIS 9 sites? Actually, if you give or take a few species, similar faunas were recovered at Barling and Cudmore Green, Essex which date to this period (Bridgeland *et al.* 2001). Nearly identical biostratagraphic arguments based on the molluscan and pollen analysis also were suggested for dating of Nightingale Estate (Keen *et al.* 1997; Green *et al.* 2004). Coope (2001) has used similar arguments to separate a series of faunas from MIS 5 and 7.

Of course the irony is that although this method is labelled 'old' and 'unscientific', stratigraphic analysis and biostratigraphy can, in certain situations, be as reliable as, if not more reliable than, 'science-based' dating.

THE EARLIEST INSECTS FROM LONDON: 350,000 TO 11,500 BP

This section outlines the nature of the insect faunas from the Thames Valley, and how this relates to its past climate and landscape in the period between 350,000 and 11,500 BP (9500 cal. PC). Figure 2.1 shows the location of the sites discussed. A more general discussion of the sequence for the whole of British Isles has been presented in Elias (2010) and in Coope (2010).

350,000 – 320,000 BP

In terms of pure chronology the oldest insect fauna recovered from Greater London is that from Nightingale Estate, Hackney. The dating and analysis of this deposit has been discussed above, along with the nature of the insect fauna.

The insect fauna clearly indicates climate conditions and the make-up of the surrounding landscape during the 'Purfleet Interglacial', approximately 320,000 years ago, if the dating is to be believed. Although no human remains were found at this site, humans have occupied the Thames Valley since at least the Anglian and we do know that humans were present in Britain since at least 700,000 years ago (Parfitt *et al.* 2005). Whatever the nature of human occupation within the Thames Valley at this time, we have solid evidence from Nightingale Estate (Green *et al.* 2004), and Barling, Essex (Bridgeland *et al.*2001) for an equitable interglacial climate, quite similar to that of today. The palaeoenvironmental remains from both sites suggest that the Lower Thames Valley was covered in a tapestry of deciduous forest, open glades and marshes.

130,000 – 120,000 BP

The next insect faunas to pick up the story of the Thames Valley are from deposits beneath Trafalgar Square. The importance of these deposits to the history of palaeoentomology has been discussed above already. The organic peats sampled date from 130,000 to 120,000 years ago (Coope 2001). These were deposited during the last interglacial (the Ipswichian Interglacial – 130,000 – 80,000 – MIS 5e). The gravels contained the bones of elephants, hippopotami, rhinoceros, wild oxen, reed deer, fallow deer and lions. In the 1950s this was the first insight into the lost world of the last interglacial for many Londoners. One particular attraction was the idea that Trafalgar Square was once home to the types of animals that now frequent the African plains. Something of the surprise and wonder generated from of this reconstruction can be seen by examining the article by Franks *et al.* in the 1958 edition of the *Illustrated London News* along with a drawing showing a range of large animals living uncomfortably close to each other, particularly carnivores and prey. However, no remains of humans or stone tools have been found at this site, or indeed, at any other sites of a similar age in Britain (Ashton *et al.* 2006).

It is thought that Britain was an island at this time, rather effectively preventing the *Homo erectus* population from reaching the Thames Valley (Ashton *et al.* 2006). Despite its obvious importance, the insect fauna from the Trafalgar Square site was never fully published. It is clear from the plant macrofossils (Franks 1960) and the mollusc faunas (Preece 1999) recovered that we see the re-establishment of temperate forest by this time, consisting of oak, hazel and maple (*Acer* spp.).

80,000 – 13,000 BP

After the Ipswichian Interglacial, north Western Europe entered a prolonged glacial cold phase, which is known as the Devensian or 'last' glaciation (MIS 5d-2). This dates from *ca.* 80,000 – 15,000 years ago. In the first part of this glacial phase, Europe was home to the Neanderthals. The later stage (from c. 40,000 years ago onwards) sees the spread of modern humanity and development of the Upper Palaeolithic way of life. There is abundant evidence from elsewhere in Britain for the

Conclusions reached from the insects present	Significant insects	Ecological data these species provide
Small shallow pools	*Agabus congener, Agabus arcticus, Colymbetes dolabratus, Ochthebius lenensis, Helophorus obscurellus, H. splendidus, H. sibericus, H. orientalis*	These species of water beetle are normally associated with shallow small pools often in glacial areas.
Marshy areas with sedges	*Notaris aethiops, Notaris bimaculatus, Grypus equiseti*	The first two of these weevils are associated with sedges and bur-reeds. The last is associated with horsetails. These plants are common in marshy areas
The landscape is sparse with little vegetation	-	There are very few phytophage insects such as the 'leaf' beetles and the weevils.
The presence of sandy and gravely bare ground	*Bembidion bipunctatum, B. dauricum, B. fellmanni, B. hasti, B. hyperboraeirum, Notiophilus aquaticus, Amara quenseli*	All of these species of 'ground beetle' prefer sandy and gravely soils. Many also like open conditions.
	Pycnoglypta lurida, Eucnecosum brachypterum, Boreaphilus henningianus, Platystethus cornutus and P. nitens, Bledius species	These 'rove beetles' favour wet moss and decaying plant matter. Often sandy substrates are also needed.
Mossy areas	*Simplocaria metallica, S. semistriata*	These beetles feed only on moss and occur in high numbers.
Animal dung	*Geotrupes, Onthophagus, Aphodius*	These are 'dung beetles' which feed on the pats of large herbivores
Artic conditions	*Bembidion lapponicum, B. dauricum, B. fellmanni, B. hasti, B. hyperboraeirum, Colymbetes dolabratus, Helophorus obscurellus, H. splendidus, H. sibericus, H. orientalis, Pycnoglypta lurida, Boreaphilus henningianus, Tachinus caelatus, Simplocaria metallica, Anthicus ater*	These species are not found in the British Isles today. Many are now only found in arctic conditions in Fenno-Scandinavia or Siberia and appear to be cold adapted or tolerant.
	Bembidion virens, Patrobus septentrionis, Amara quenseli, Agabus arcticus, Agabus congener, Ochthebius lenensis, Eucnecosum brachypterum, Notaris aethiops	These are all species that today are limited to the north of Scotland, often in mountainous areas. Their geographical range extends into the artic areas of northern Europe.

Figure 2.6. The ecology of the significant insects from the lower deposits at South Kensington and their interpretation

periodic occupation of the British Isles by humans throughout this time (Ashton *et al.* 2006). However, in the Lower Thames Valley, the evidence for the presence of Middle and Upper Palaeolithic populations is very patchy (Merriman 1990). The impression gained is of people moving through the valley either following the herds of animals on their annual migration or simply following the Thames Valley corridor on their way to somewhere else.

The earliest deposits sampled in the area of London from the Ipswichian come from South Kensington (Coope *et al.* 1997). These were found during the construction of the Ismaili Centre opposite the Victoria and Albert Museum. This deposit is radiocarbon dated to before 45,000 years ago. Russell Coope's insect faunas from this deposit are very different to those described for earlier periods. They clearly come from a time when the Lower Thames Valley was subject to much more severe climatic conditions than the benign ones described for the preceding inter-glacial periods. Full glacial conditions seem to prevail, which means the deposit probably dates from MIS 4.

Once again the landscape reconstruction offered by the insects is quite detailed (Coope e*t al.* 1997). A general impression of the fauna is described here, but a more detailed summary is presented in Figure 2.6, which also indicates how various species of beetle support these conclusions. The insects suggest that small shallow pools with areas of sedges (*Carex*) were located in a relatively barren, marshy landscape. There is evidence for a few isolated patches of thin grassy vegetation and mossy ground. Open areas of gravel and sand also occurred. The 'dung beetles' present suggest that large herbivores grazed across the area. In terms of climatic indicators there are a number of beetles present that are extinct or have northern distributions in the British Isles today. Particularly good examples of this being the trio of omaliid 'rove beetles' – *Pycnoglypta lurida, Eucnecosum brachypterum,* and *Boreaphilus henningianus.* These three species are always common components of glacial faunas, leading Osborne to label Ipswichian Ice Age deposits as the 'omalinine desert' (*pers. com.*).

However, the site in South Kensington has produced perhaps the most spectacular examples of 'indicator species' with the recovery of two small water beetles. The nearest place to the Thames Valley that *Helophorus obscurellus* occurs today is the Kanin Peninsula in arctic Russia (see Figure 2.7). An even more 'exotic species' is *Helophorus splendidus* which today occurs in arctic Siberia eastwards of the gulf of the River Ob (see Figure 2.7).

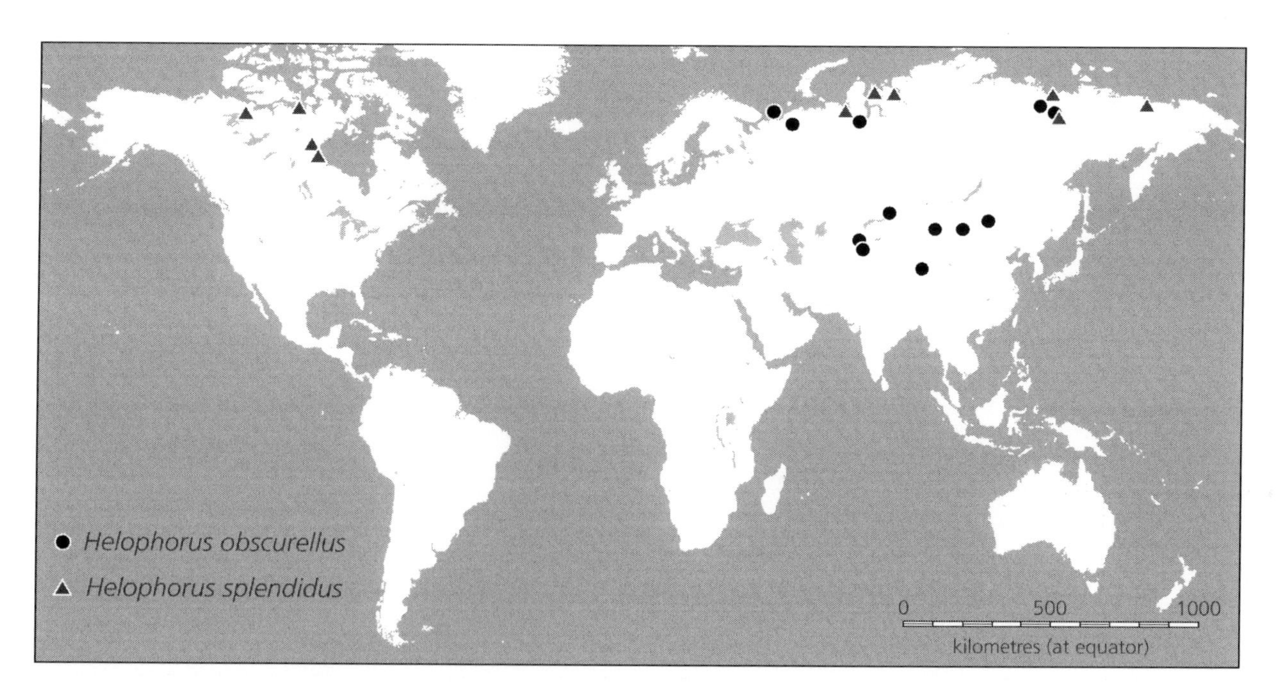

Figure 2.7 The modern distribution of *Helophorus osbcurellus* and *Helophorus splendidus* (redrawn from Coope and Brophy 1972)

Of course, what should be obvious by now is that this is an insect fauna typical of glacial conditions. Using many of these cold adapted species Coope was able to suggest that the mean temperature in July would have been 9°C (the equivalent of a mild winter's day today) and the mean temperature in the middle of the winter a perishing –22°C. Of course, this is the mean temperature and conditions could be considerably colder at times. The sediments from directly above this clearly glacial material date from around 43,000 years ago (Coope *et al.* 1997). The insect fauna recovered is very different from those derived from the earlier, underlying glacial deposits. In fact, many of the insect species recovered from these upper layers are identical to those found in the Nightingale Estate deposits. Again, there is evidence for a fast flowing, gravely river with areas of slower-flowing water where sedges and reeds occur in places. Beyond the river was grassland full of meadow plants and, given the large number of dung beetles recovered, grazed by large herbivores. Deposits from Isleworth, West London (Figure 2.1), with the same date essentially contained a similar insect fauna (Coope and Angus 1975; Kerney *et al.* 1982). Isleworth also produced the skeletal remains of bison and reindeer, possibly suggesting the identity of the large herbivores at South Kensington. Again, Coope used the beetles present to suggest the seasonal temperature range at both sites (Coope and Angus 1975; Kerney *et al.* 1982; Coope *et al.* 1997).

During in the Thames Valley maximum mean summer temperatures were a rather pleasant 17°C with the coldest part of the winter a relatively comfortable but frosty mean temperature of –3.5°C, so cool rather than perishing. However, unlike the preceding warm periods there is no evidence for the presence of trees or forests. Initially this absence resulted in some confusion (Coope and Angus 1975). However, with better dating, and information from the oxygen isotope record, we now understand why the trees are absent. These sites belong to the earliest phase of a common 'geological' deposit often found in gravel quarry sequences. This period first described by Professor Shotton from Birmingham in the 1950s and named by him as the Upton Warren 'interstadial' phase. This is a 20,000-year long period of cool and continental conditions sandwiched between two extreme glacial periods, now known to start with a 500 year long period of temperatures similar to today (Atkinson *et al.* 1987). This is probably too short a period of time for trees to fully re-colonise northern Europe, or the Thames Valley for that matter, from their glacial refuges in southern Europe (Coope and Angus 1975; Coope *et al.* 1997). Moreover, this is also too short a period of time for the soil development needed to support dense stands of trees (Coope and Angus 1975; Coope *et al.* 1997).

Later deposits at Kempton Park, Sunbury (Gibbard *et al.* 1982) show that the warm, but short period at *ca.* 43,000 years ago was followed by a slow decline towards full glaciation. The lower deposits at Kempton Park contain some temperate insects, but by 35,000 years ago there is evidence for a return of cold adapted species, for example, *Diacheila polita, Bembidion dauricum, B. hasti, Colymbetes dolabratus, Helophorus osbcurellus, H. jacutus and Eucnecosum brachypterum, Helophorus aspericollis, Tachinus jacuticus* and *Aphodius jacobsoni* which all have modern distributions that include Siberia and Mongolia, suggesting that continental conditions existed (Gibbard *et al.* 1982). Unfortunately, Coope was unable to use the MCR method to give a numerical estimate of temperatures at the time since the insect faunas recovered were small and the sampling interval had been particularly coarse (Gibbard *et al.* 1982).

For the next 30,000 years there are no insect faunas from the Lower Thames Valley and very few pollen diagrams. Elsewhere in Britain and Northern Europe this period sees the return of very intense glacial conditions at around 22,000 years ago (Atkinson *et al.* 1987). Humans probably abandoned Britain during the period and only returned with the warming of the environment at the start of our present interglacial, the Holocene starting at *ca.* 9,500 years ago (Ashton *et al.* 2006).

In terms of pollen profiles from the Thames Valley this period is only really represented by deposits from Silvertown, Newham, Bramcote Green, Bermondsey and Three Ways Wharf, Uxbridge that date to the last interstadial of the Devensian (the Windermere interstadial) (Rackham and Sidell 2000). This is a dramatic, but very short, warm phase dating between *ca.* 13,500 and 11,000 cal BC (Atkinson *et al.* 1987; Lowe *et al.* 1995). Pollen analysis at these sites suggests the presence of species rich meadow dominated by sedge and grasses. At Bramcote Green there is limited evidence for willow, birch and juniper in the area, but not for forest or substantial tree cover (Thomas and Rackham 1996). This phase is followed by a decline into extreme artic conditions that last for around 1000 years (the Loch Lomond Stadial) (Atkinson *et al.* 1987; Lowe *et al.* 1995).

CONCLUSION

This chapter reviewed our present knowledge of how the Thames valley changed between 350,000 – 11,500 years BP (9500 cal BC) and how insect faunas have contributed to that understanding. Hopefully, it has also shown how powerful a technique palaeoentomology is when used in reconstructions of climate and landscape.

One thing that is clear is that despite the detail of the story told there are many long gaps in the record. Particularly noticeable is the complete absence of faunas from between 22,000 and 9500 years ago (in fact this absence extends down to *ca.* 7000 cal. BC). Some periods are only represented by a single site. There is clearly much more work that could be done, particularly as modern development occurs and deposits such as this are exposed. It also needs to be remembered that throughout much of this period the Thames Valley was not on the edge of Britain, and Britain was not on the edge of Europe. During this phase this area was part of a larger north European plain and work undertaken in the Thames Valley has clear European importance.

CHAPTER 3: HOLOCENE INSECTS FROM THE THAMES VALLEY: FROM THE MESOLITHIC TO THE ARRIVAL OF THE ROMANS (9500 cal BC – cal AD 50)

This chapter outlines how Thames Valley insect faunas can help us understand how the landscape changed during the Holocene. It also demonstrates how insect remains can be used to try to shed light on a number of research issues concerning how ancient Britains used and changed the landscape before the arrival of the Romans and the foundation of the City. The sites mentioned in the text are illustrated in Figure 3.1.

THE EARLY POSTGLACIAL (MESOLITHIC) *ca.* 9500 TO 4500 cal BC

We now have a clear understanding of how the climate of Northern Europe and the British Isles changed as glacial conditions eased at around 9,500 BC. This is established from arctic ice core records (Dansgaard *et al.* 1989; Lowe *et al.* 1995; Lowe and Walker 1997a, 1997b) and from Coope's reconstruction of temperatures using insect remains (Coope and Brophy 1972; Atkinson *et al.* 1987, Elias 2010). Coope's temperature curve for this period was shown in Figure 2.4. By 9500 cal BC there was a dramatic rise in summer temperatures by as much as 20°C and winter temperatures thawed by an even greater margin. It is also clear that this rapid change in climate actually occurred over a very short period of time, perhaps as much a 9°C over 50 years in terms of sea surface temperature (Bell and Walker 2005). The reason for this rapid change has been debated for the last 20 years but the consensus view suggests that it is linked to the re-establishment of the Gulf Stream, and therefore, the return of warm waters to the north Atlantic (Dansgaard *et al.* 1989).

Pollen analysis over the last 50 years has also allowed us to reconstruct an detailed picture of environment and landscape change in the period between 9500 and 7000 BC in many regions of the British Isles. Unfortunately, the specific picture for Central London and the Lower Thames Valley is less complete, though there has recently been a review of this data from the Surrey side of the Thames by Branch and Green (2004) and a limited survey for central London by Rackham and Sidell (2000). For such a large geographical area there is again a surprisingly limited range of sites studied, many of which have only been published in the last decade (Lewis 2000a; Rackham and Sidell 2000). These include the peat bed at Enfield Lock, Middlesex (Chambers *et al.* 1996), the peat beds and channel fills at Bramcote Green, Bermondsey (Thomas and Rackham 1996), the complete sequence of sediments from the boreholes at West Silvertown, Newham (Wilkinson *et al.* 2000), the early Mesolithic sites at Three Ways Wharf, Uxbridge (Lewis *et al.* 1992) and a series of sites associated with the Thames at Staines and Meadlake, Surrey (Branch and Green 2004).

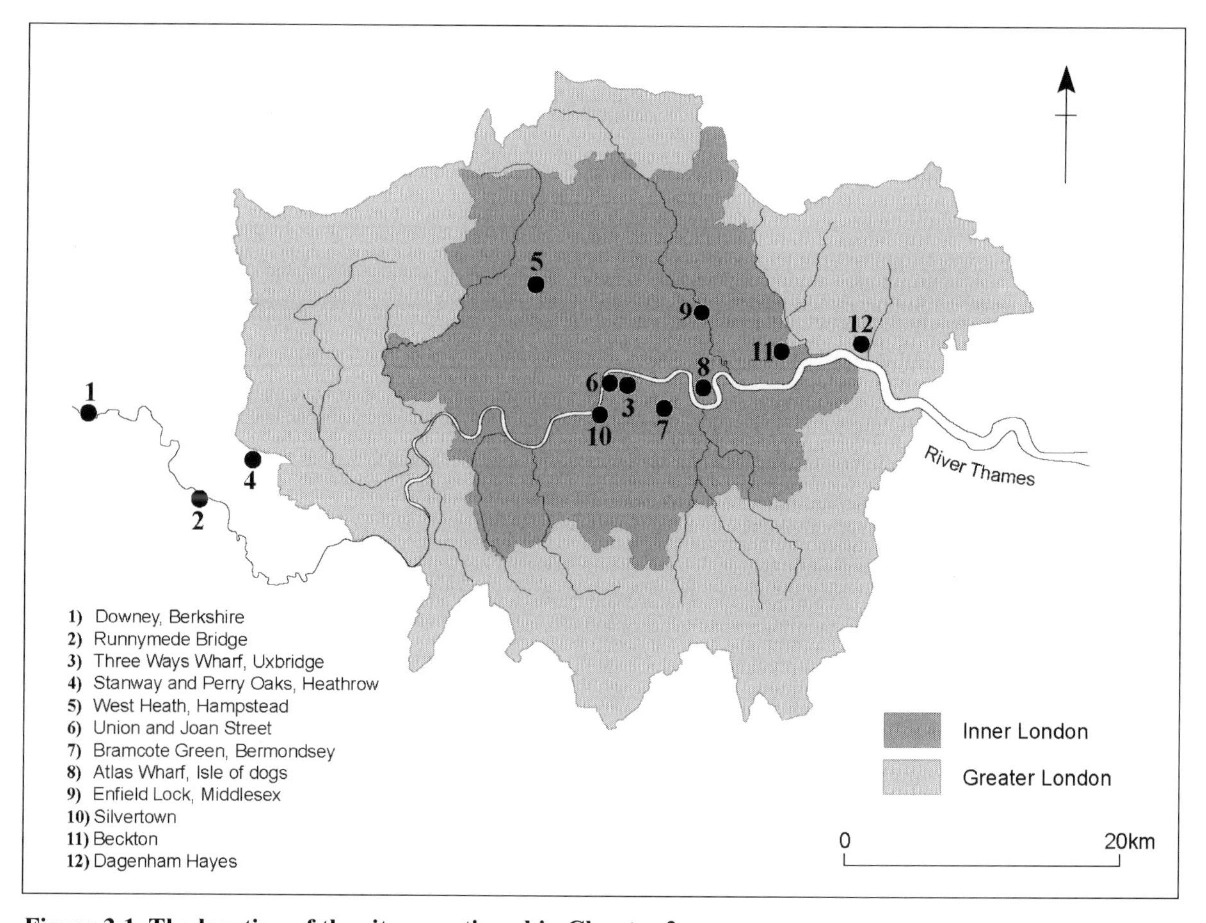

Figure 3.1. The location of the sites mentioned in Chapter 3.

At Silvertown (Wilkinson *et al.* 2000) the pollen evidence suggests a marked change at around 9500 cal BC with pine and birch becoming dominant in the pollen record. This is also seen in the lowest and contemporaneous sediments at Enfield Lock (Chambers *et al.* 1996), Bramcote Green (Thomas and Rackham 1996) and the sites at Staines and Meadlake (Branch and Green 2004). Though not dated by radiocarbon determination, the lowest deposits at Three Ways Wharf are believed to be roughly from the same period (Lewis *et al.* 1992). The dominance of pine seems to decrease around 8000 BC at all sites. After this, most sites see the gradual establishment of mid-Holocene 'climax' woodland. The sequence of tree colonisation across the terraces of the Thames Valley is best seen at Silvertown, Bramcote Green, Enfield Lock and Staines and Meadlake where hazel, oak and, to a lesser extent, elm (*Ulmus*) become dominant at around 7000 BC. Lime (*Tillia*) and alder appear to become dominant features of this mixed deciduous woodland by around 6400 BC. The later part of this succession, particularly the growing importance of lime, is also seen at sites in Bermondsey, Southwark and Westminster along the route of the Jubilee Line Extension (Sidell *et al.* 2000). A similar pattern of colonisation by tree species is also seen in the pollen sequences from up river at Runnymede Bridge, Surrey (Scaife 2000).

One curious aspect of the sequences from the small tributary sites to the north and west of London at Three Ways Wharf and Enfield Lock is that they contain abundant charcoal. Micro-charcoal, effectively the remains of smoke, occurs in the pollen slides from both sites (Lewis *et al.* 1992; Chambers *et al.* 1996). There are at least three distinct periods of burning at both sites and slight indications in the pollen sequences that this affected the composition of the trees in the surrounding forests (Lewis *et al.* 1992; Chambers *et al.* 1996). There has been a long running argument in the archaeological literature as to whether such burning was deliberate, accidental or the result of natural forest fires (e.g. summary in Bell and Walker 2005; Bell 2008; Gaffney *et al.* 2009). It has been suggested that burning may clear undergrowth for hunting and increase the amount of browse for wild deer populations (Simmonds 1975, 1996; Mellars 1976; Innes and Simmonds 1988). Certainly the scatters of Mesolithic flints at Uxbridge, West Heath Spa and Three Ways Wharf (Merriman 1990) and at the Stanway site, Heathrow (Lewis *et al.* 2006) indicate that postglacial hunters and gathers were present in the area. In terms of the area of the modern city the only substantial find dating to the Mesolithic are the three flint scatters and a number of hearths at the B&Q site in South Bermondsey, at Marlborough Grove and at Waterloo (Lewis 2000a; Sidell *et al.* 2002; Sidell *pers. com.*). These seem to have been used by people exploiting the area around 'Bermondsey Lake' (the remains of a postglacial 'cut' of the River Thames) for hunting and gathering during the early Mesolithic (Lewis 2000a; Sidell *et al.* 2002).

This leads us on to consider the rather attractive story of Mesolithic peoples deliberately altering areas of woodland and starting the process of 'nascent herding', seen as the early foundations for the adoption of agriculture (e.g. Simmonds and Dimbleby 1974; Simmonds 1975; Zvelebil 1994). However, this theory has recently been questioned (Bennett *et al.* 1990; Huntley 1993; Bell 2008). One alternative is that this burning results from natural forest fires during periods of dry climatic conditions with Mesolithic people taking advantage of the clearance (Tipping 1996, 2004). However, it is possible that the Mesolithic occupation of the Thames tributaries was quite intensive. The 'burning' could actually result from the repeated use of small fires at campsites and living areas over long periods of time (Chambers *et al.* 1996). If this latter theory is the case, it suggests that the use of some areas of the Thames Valley may have been more intensive than the sparse finds of worked flint from this period would suggest.

Though of course this is special pleading; it is a shame that no insect faunas were recovered from any of these sites. With hindsight it is clear that archaeoentomological results could have played a valuable role in both landscape and land use reconstruction. Certainly a number of the research questions and issues from all of the sites could have been addressed. In fact, I looked at a few insect faunas from the early periods at Bramcote Green right at the start of my archaeoentomological career. However, the size of sample collected for archaeoentomological analysis at the site was small, often only a litre or two. This meant that the faunas were often of little interpretative value. Another lesson to be learnt by all; sample size should be a minimum of 10 litres at least for insect remains. There is nothing more frustrating for an archaeoentomologist than if the recovered insects are of little use because the volume of material sampled was below the recommended amount.

There are very few sites with Mesolithic insect faunas from the Thames Valley south of the Goring Gap. Perhaps the clearest sequence can be seen at the site of Runnymede Bridge (Robinson 2000a). Though strictly located outside of the area of Greater London, it is discussed here for the sake of completeness. A cynic could also argue, given this apparent 'blank period' for insect remains from London, that the Runnymede results may have been included out of desperation. Despite the small samples sizes collected (Robinson 2000a, 150), the Runnymede beetle faunas do suggest some general trends for the period between 9000 and 4500 cal. BC. The water beetles suggest that this section of Thames flowed over clean gravel throughout the entirety of this period. The evidence for this comes from the same range of the elmid 'riffle beetles' discussed above for interglacial deposits. Both *Macronychus quadrituberculatus* and *Stenelmis canaliculata* also were recovered at Runnymede Bridge (Robinson 2000a). These species are thought to be particularly 'fussy' and need oxygenated and silt free waters (Osborne 1988; Smith 2000b) and are today absent from the Thames valley (Robinson 2000a). It has been suggested that the 'riffle beetles' were common in all British river systems before the start of large-scale erosion and the deposition of clay alluvium from the Early Iron Age onwards (c. 800 cal BC) (Osborne 1988; Smith 2000b; Smith and Howard 2004) and relatively uncommon after this date. Recent work by Scot Elias also suggests that *Stenelmis consobrina* was also present in the Thames during this period (Elias *et al.* 2009). This is a striking result given that this species was last seen in the Thames Basin during the Ipswichian interglacial (Franks *et al.* 1958). A wider investigation of the elmid faunas of The Lower Thames Valley from the Later Holocene, particularly from after the Iron Age, could help establish the nature and extent of erosion and alluviation in this river system. Unfortunately, as will be seen below, riverine deposits of this date are quite rare. How the pattern of river development may have played out to the east of the modern city is unclear, but it was probably also affected by raised water tables and sea level rise at this time (Devoy 1979).

Robinson (2000a) suggests that the presence of the leaf beetles *Phyllodecta* and *Chalcoides* from the early Mesolithic deposits at Runnymede indicate that willows dominated the riverbanks at this time. The Later Mesolithic deposits (<6000 cal BC), however, contain insects such as the attractive purple 'leaf beetle' *Agelastica alni*, which are associated with the foliage of alder. A similar pattern is also seen in the insect fauna from Dorney (Parker and Robinson 2003). This pattern also is reflected in the pollen (Scaife 2000) and plant macrofossil (Robinson 2000a) results from Runnymede. One sample (sample 40) from later Mesolithic deposits at Runnymede enables us to see the nature of the woodland growing in the adjacent gavel terraces. A number of individuals of scolytid 'bark beetles' (*Scolytus scolytus* and *Hylesinus oleiperda*) recovered are associated with elm (*Ulmus*) and ash (*Fraxinus*). Several species of 'cossinine' weevil that are associated with dead wood in mature forests were recovered as well. Very few insects from this sample could be considered to be associated with open ground, forest glades or animal dung leading Robinson (2000a) to conclude that the woodland was, at least at this location, dense and enclosed. A number of other Mesolithic faunas have also been recovered from the isolated tree throw hole at Dorney, Berkshire (Parker and Robinson 2003), the Horton Kirby Paper Mill, South Darenth, Kent (Elias *et al.* in press) and Crossness, Bexley (Elias *et al.* 2009). Though the faunas are often limited in size they to appear to generally support the pattern established at Runnymede.

More will be said about the difficulty of reconstructing the nature of early woodland in Chapter 4.

THE NEOLITHIC THAMES VALLEY AND ITS INSECTS

The Neolithic sites mentioned in the text are shown in Figure 3.1. The Neolithic often is held to be the period of the first introduction into the British Isles of farming using domesticated livestock and arable plants. Along with this farming practice came pottery, ground stone axes and the other parts of the 'Neolithic package'. Classically this was thought to start at around 4,000 cal BC in Britain and arrived with a sudden 'bang'. It was also thought that hunting and gathering as a way of life was quickly replaced by farming (though this is perhaps a bit unfair to individual authors see Helbaek 1975; Mercer 1981; Darvill 1987). However, there is an increasing range of evidence that suggests that the uptake of the 'Neolithic package' may have been much slower and to have actually started around 4500 cal BC in some places in Britain. It is now thought that for many Neolithic communities agriculture may have been a minor part of subsistence, with hunting and gathering remaining dominant (Moffett *et al.* 1989; Edmonds 1995; Whittle 1997; Evans *et al.* 1999; Richmond 1999; J. Thomas, 1999, 2008; Robinson 2000c, Bell and Walker 2005; Bradley 2006) or that there was a large degree of regional variation in subsistence practice (Pollard 1999, 2000, 2004; Jones 2000; J. Thomas 2003, Jones and Rowley-Conwy 2007).

In terms of the Thames Valley there seems to have been limited 'Neolithic' activity at the start of the period (Merriman 1990; Lewis 2000a, 2000b; Cotton 2000, 2004) with finds of early Neolithic pottery and stone tools comparatively rare (Sidell *et al.* 2002; Cotton 2004), though recent research at Horton Quarry, near Heathrow may suggest a concentration of activity in this area (Alistair Barkley *pers. comm.*). This apparent absence could, of course, be a direct reflection of the level of human activity at the time (Cotton 2000). In addition, the difficulty of prospecting for early sites in river floodplains in general (Howard and Macklin 1999) and beneath the City of London in particular, has been widely recognised (Merriman 1990; Sidell *et al.* 2000, 2002; Lewis 2000b). Moreover, there is clear evidence that the area to the east of the modern city was progressively affected by rising water tables and sea level change in this period. This results in the formation of large areas of alder carr and estuary flats (Devoy 1979, 1980), environments not conducive to early agriculture.

In terms of the Middle Neolithic the only location that is currently known to have had substantial activity is the riverside at Runnymede Bridge, where there are various brushwood structures dating from between 4000 and 3500 cal BC (Needham 1991). Away from the floodplain, artefact evidence for the Middle Neolithic is extremely ephemeral (Lewis 2000b). By the Middle and Late Neolithic (4000 – 2000 BC) there is a scatter of small isolated 'monuments' such as the causeway enclosure at Yeoveny Lodge, Staines, the large Stanwell cursus and enclosure at Heathrow (Cotton 2004; Lewis 2000b; Lewis *et al.* 2006) and a scattered number of 'long mortuary enclosures' (Cotton 2000, 2004). Further away other isolated small ditch 'monuments' and 'pit diggings' seem to dominate (Lewis 2000b; Cotton 2000). There appears to have been no established tradition for the construction of Middle Neolithic long barrows or Later Neolithic henges in the area (Merriman 1990; Lewis 2000b; Cotton 2000, 2004). Given that this is such a large area of southern Britain, it is striking how limited Neolithic remains appear to be in the Thames Valley

(Lewis 2000b). This is particularly apparent when the distribution of Neolithic sites is compared to that of the contemporaneous Upper Thames Valley or the downlands of Wessex. When discussing the 'Neolithic settlement' of the Lower Thames Valley Cotton (2000, 25) suggests *"perhaps the key word is 'mobility', a concept that runs counter to the traditional view of 'sedentary farmers'."*

Although Neolithic sites are limited, the palaeoenvironmental results from a number of sedimentary deposits do at least allow us to reconstruct the landscape of the Thames valley during this period. The pollen profile from Silvertown (Wilkinson *et al.* 2000) clearly indicates that by the start of the Middle Neolithic (*c.* 3900 cal. BC) the terrace woodlands had started to change. In particular the proportions of elm, oak and lime present decline in favour of open ground species, with a notable expansion in grasses and bracken. Though the interpretation of percentage pollen diagrams has to be treated with some scepticism, there does appear to be evidence for small and temporary clearings around the start of the Middle Neolithic. Similar indicators for small-scale clearings at this time are also seen in the pollen profiles at West Heath Spa, Hampstead (Greig 1989) Union and Joan Street in the City (Sidell *et al.* 2000), and at the Stanwell Circus (Lewis *et al.* 2006). At Silvertown, Union Street and Bryan Road (Sidell *et al.* 2002) possible cereal and plantain pollen, both classic indicators for human activity, are also recorded (Wilkinson *et al.* 2000; Sidell *et al.* 2000). However, in the middle Thames Valley in the area centred on Staines there seems to be only limited evidence for cultivation and clearance at this time (Branch and Green 2004).

Do we see similar patterns and hints of human activity in the insect faunas from the sites from the lower Thames that date to this period? Certainly, at both West Heath Spa, Hampstead Heath (Girling 1989a) and at Runnymede Bridge (Robinson 2000a) the insect remains include a large number of tree-dependant species as do the five Neolithic sites published in the recent survey by Elias (Elias *et al.* 2009). This appears to indicate the continued presence of mixed deciduous woodland, with no evidence for substantial clearings for agriculture. At Runnymede Bridge, species associated with woodland account for about 10% of the terrestrial fauna (Robinson 2000a) and at West Heath Spa around 30% (Girling 1989a; Whitehouse and Smith 2010). Though at first sight, the relative proportion of the tree-dependant taxa may seem low, it is in line with a number of other Early Holocene sites both before and after the start of the Neolithic (Robinson 2000a; 2000b; Whitehouse and Smith 2004; Whitehouse and Smith 2010). This is demonstrated in Figure 3.2 where the proportions of woodland associated species of beetle from a wide range of sites are shown (Whitehouse and Smith 2010). The explanation for this is that the terrestrial beetle fauna from an archaeological site will always contain a large proportion of species that do not have any particular ecological preference. This means if 30% of the taxa are associated with a specific habitat it was a fairly dominant aspect of the wider environment. Robinson (1991; 2000b) has argued that any figure over 20% probably indicates dense, continuous canopy woodland. A range of modern analogue studies has recently suggested that Robinson's figure is essentially correct (Kenward 2006; Smith *et al.* 2010). The issue of whether this is really the case, and the meaning of various percentages of ecological groupings, will be discussed further in Chapter 4.

At West Heath Spa, however, there is a slight suggestion of possible human activity in the environmental evidence recovered. The pollen diagrams from this site appear to contain evidence for the presence of clearings after the elm decline, at some point between *ca.* 3900 – 3700 BC. The percentage of tree species drops to 12% in the deposits just before the elm decline and eventually declines to 2% at the top of the section (Greig 1989).

By the start of the Middle Neolithic, Girling (1989a) suggests that the insects provide more evidence for large herbivores, possibly domestic stock, in the area. This is particularly suggested by an increase in the number of dung beetles recovered. These now represent as much as 5% of the terrestrial fauna. A similar situation is seen at Runnymede Bridge, where the Early - Middle Neolithic insect faunas also contained a relatively large proportion of dung beetles (Robinson 2000a). This, and a small number of insect species indicative of grassland, led Robinson (2000a, 151) to conclude that there must have been a concentration of large herbivores in the area from the start of the Neolithic. He suggests that these were domesticated cattle. Robinson also notes that there are no plant macrofossils that indicate the presence of open spaces along the riverbank. This is not what you would expect if the dung beetles were associated with wild animals that were only coming down to open areas of riverbank for access to water. He holds, therefore, that this probably represents deliberate herding in the woodlands on the gravel terraces. By the Middle Neolithic at Runnymede Bridge (*ca.* 3500 cal BC onwards) there appears to be even larger proportions of indicators for open ground and grazing (around 9%) in the insect fauna (Robinson 1991). At the same time, the proportion of insect species associated with woodland drops to just over 7%, suggesting a progressive clearance in the area. Girling (1989a) also suggests that the data from West Heath Spa indicates that by the Middle Neolithic domestic grazing by cattle, and probably pig, was occurring in the Neolithic forest. However, there are some theoretical problems with this interpretation, and these will be addressed in Chapter 4 when the difficulties of interpreting this type of fauna are discussed. One issue to note about the sequence from West Heath Spa is that the dating is extremely poor. There are no radiocarbon dates on the material and all of the dates discussed above are extrapolated directly from the pollen. To some extent the chronology presented above remains speculative.

Another site from this period that provides an insight into the nature of woodland during the later Neolithic is Atlas Wharf, Isle of Dogs (Smith 1999). At Atlas Wharf there is clear evidence for alder woodland on the floodplain

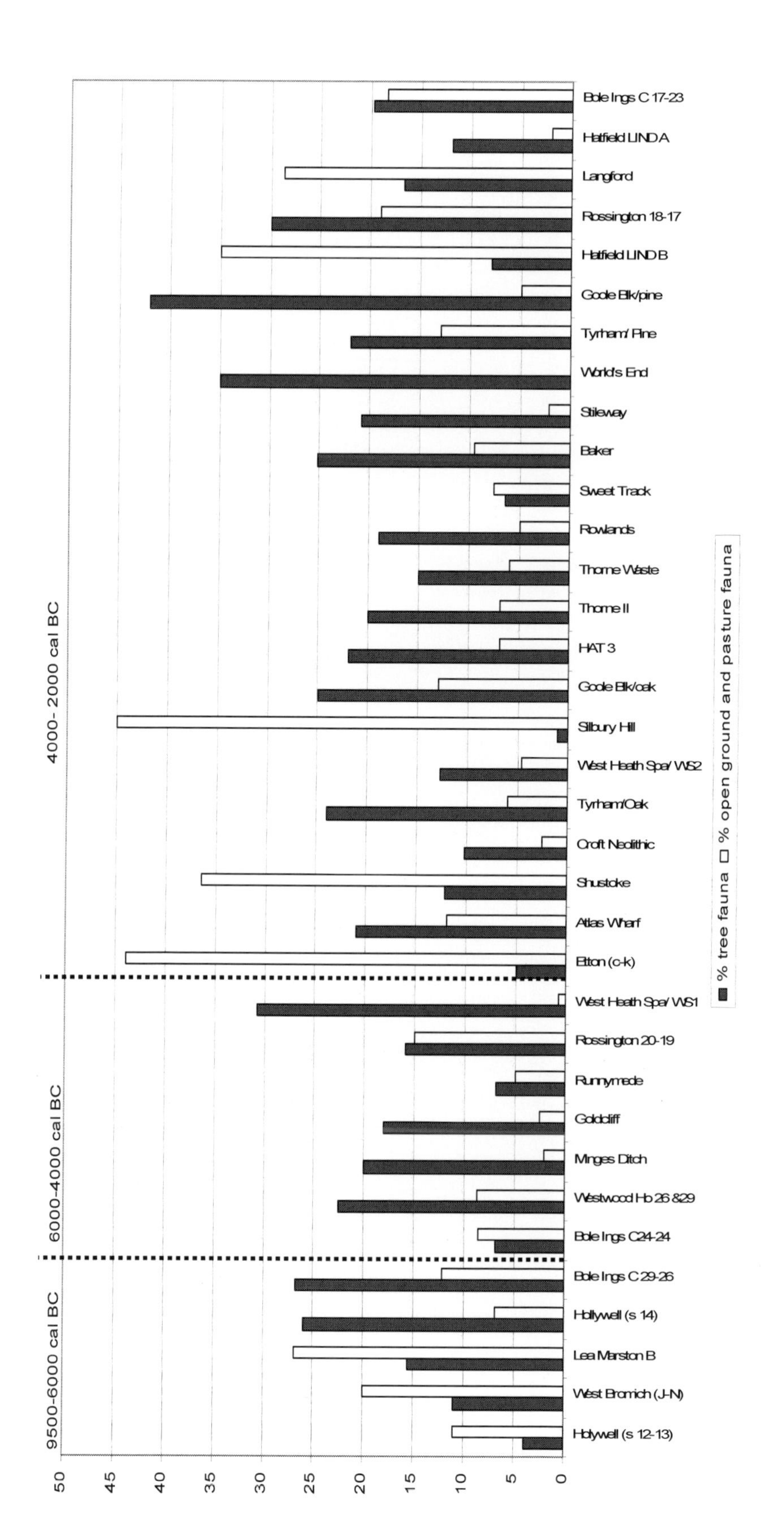

Figure 3.2. **The proportion of beetles that indicate woodland and clearance from a number of early Holocene sites in the UK (Whitehouse and Smith 2010).**

throughout the sequence. Three species of beetle (*Melasoma aenea, Agelastica alni* and *Dryocoaetes alni*) indicate the presence of old and mature alder carr. Away from the floodplain beech (with which *Rhysodes sulcatus* and *Melasis buprestoides* are associated), ash (associated with *Hylesinus crenatus, H. oleiperda* and *leperisinus varius*), elm (indicated by *Pteleobius vittatus*) and lime (indicated by *Ernoporus caucasicus*) appear to have dominated woodland on the Isle of Dogs at this time. Similar to the pattern seen at West Heath Spa and Runnymede there are also small numbers of dung beetles and other indicators for open grassy ground perhaps suggesting the timber structures might have had some role in the movement of cattle. However, the grazing of cattle in the Neolithic Thames basin may not have been widespread since the recent survey of 5 Neolithic sites by Elias (Elias *et al.* 2009) does not show this pattern or particularly high numbers of dung beetles.

All the Neolithic sites from the Thames Valley also produce two species of scolytid 'bark beetles', which should be discussed in detail since both of these species have now become the centre of two specific archaeological arguments.

The 'shot borer', *Ernoporus caucasicus,* is only associated with lime. Today it is virtually extinct in the British Isles with its modern distribution limited to a number of old 'specimen' trees in Windsor Great Park, Berkshire and Moccas Park, Hereford (Hyman and Parsons 1992). However, it clearly had a much wider distribution in the past. It is commonly recovered in lowland insect faunas studied from this period including Runnymede Bridge (Robinson 1991), various sites in the Upper Thames Valley (Robinson 1993) and the Trent Valley (Greenwood and Smith 2005). It clearly supports Greig's (1982) suggestion that lime trees, despite the rarity of their pollen in cores of this date, were an important component of early Holocene woodland. Another species from West Heath that has gained some notoriety is the 'elm bark' beetle *Scolytus scolytus* (the joke in entomological circles is 'so nice they named it twice'). This species, has gained a name for itself as a potential cause, and therefore, explanation for the Neolithic event known as the 'elm decline' (*sensu* Iversen 1941; Parker *et al.* 2002; Bell and Walker 2005). All pollen profiles, including those from Greater London, show a drop of 50–80% in the abundance of elm pollen at around 3900 cal BC. Iversen (1941), who first recorded the phenomenon, placed the blame on climatic change. Others noted the coincidental link between the time of first appearance of farming and the elm decline and suggested that it represented evidence for early clearing. This view is now discounted, mainly because there is now abundant evidence for Neolithic activity and farming before the elm decline in many regions of Northern Europe (Bell and Walker 2005). The actual number of elm trees lost from the landscape (possibly around 10% of the total tree cover) far outweighs that needed to establish a clearing for agriculture (Rowley-Conwy 1991). This is particularly true if early Neolithic farming was on a very small scale (Richmond 1999; J. Thomas 1999). To me, such a dedicated removal of only elm trees seems rather strange, and possibly a little vindictive. A rather more elegant explanation was suggested by Troels-Smith (1960) who argued that the elms were not lost from the landscape but were, along with other species of trees, seasonally cut to produce leaf fodder every other year. This practice can still be seen in some areas of southern Europe today (Rasmussen 1989; Halstead and Tierney 1998; Smith 1998). Elm flowers on branches that are at least three years old. This means that, in terms of pollen production, elm is probably more adversely affected by leaf foddering with a short cutting period than many other trees. Unfortunately, the sheer scale of activity needed to produce this degree of decline in pollen production seems unsupportable given the size of Neolithic populations. Rowley-Conwy (1981) in a nice piece of 'tongue in cheek' logic suggests that if Troels-Smith's arguments were correct the leaf fodder produced would have supported a cattle herd of several million in northern Europe. Where *Scolytus scolytus* comes into this debate is that it is the vector for the fungus *Ophiostoma (syn. Ceratocystis) ulmi* which caused Dutch elm disease in the 1960s and 1970s. It has now been widely accepted that a similar disease outbreak is a likely explanation for the rapid and widespread decline in Neolithic elm pollen (Bell and Walker 2005). West Heath Spa comes into this story because Girling found three individuals of *S. scolytus* at the site, dated to near the time of the elm decline. She drew this to the attention of the palaeoecological community in number of academic papers (Girling and Greig 1977, 1985; Girling 1989a) and explained the connection between this species of beetle and disease in elm trees. In the odd way that these things can happen, some people in the archaeological community took the evidence from West Heath Spa as a 'smoking gun' that proved the argument correct. However, as Robinson (2000b) and Clark and Edwards (2004) have pointed out this is far from being firmly established for the following reasons:

1) Though there is evidence for *Scolytus scolytus* at West Heath Spa there is no physical evidence for the fungus *Ophiostoma ulmi.*
2) *Scolytus scolytus* is a common part of the early Holocene insect fauna and its presence is not surprising or indicative of large numbers of elm trees 'in trouble'
3) In fact the specimens at West Heath Spa were recorded some 20 cm below the level of the 'elm decline' as interpreted by the pollen. Unfortunately, there is no radiocarbon dating at this site and the time interval that this represents is not clear.

So perhaps we need to take a step back and stop seeing the evidence from West Heath Spa as directly helpful to this argument. I agree that, on balance, the best explanation for the elm decline is an outbreak of disease. Personally I feel that though *Scolytus scolytus* may be a likely suspect, the evidence is circumstantial and the case is not positively proven.

INTO THE BRONZE AND IRON AGES: TRACKWAYS, FIELDS AND DUNG BEETLES

In all the pollen sequences from Greater London there is a large drop off in the proportion of lime and other trees such as oak and hazel between *ca*. 2000 and 800 cal BC (Sidell *et al.* 2000; Branch and Green 2004; Rackham and Sidell 2000). This is thought to be indicative of forest clearance in advance of the development of large-scale intensive farming in later prehistory. The date of the start of this trend in the pollen diagrams varies widely. For example, the sites from along the course of the Jubilee Line Extension indicate this event occurred at a number of dates ranging from as early as *c*. 2500 cal BC at Union Street (Sidell *et al.* 2002) to as late as *c*. 2000 cal BC at Canada Water (Sidell *et al.* 2000). At all of the Jubilee Line Extension sites this decline in tree pollen coincides with an expansion of grass pollen and other indicators for disturbed ground, such as plantains, dock and bracken. In addition, several sites produced wheat, barley or rye pollen during the Iron Age. The pollen sequence at Silvertown, in the marshlands to the east of the modern city, also sees a lime decline occurring at *ca.* 1500-1150 cal BC (the Middle Bronze Age), just before the area was inundated by rising sea levels (Wilkinson *et al.* 2000). This is significantly earlier than other dates in this area of London (Meddens 1996; C. Thomas and Rackham 1996; Sidell *et al.* 2000). A similar decline is also seen in the pollen diagrams from sites to the west of the modern city. At both Runnymede (Scaife 2000) and West Heath Spa (Greig 1989) a dramatic decline in lime and other forest trees is accompanied by corresponding rises in both herb and cereal pollen in the Bronze Age. Unfortunately, this event was not radiocarbon dated at either site. Similarly, the pollen from the Perry Oaks watering holes also indicates the establishment of a similar environment by the Middle Bronze Age (Lewis *et al.* 2006). It therefore seems that widespread and large-scale clearance for agriculture started during the Bronze Age in the Greater London region, but locally the inception date of this event could have varied widely (Branch and Green 2004).

A number of insect faunas from the Middle and Late Bronze Age periods appear to show evidence for progressive clearance. There is some isolated material of this date in the sequences at Runnymede (Robinson 2000a). Species associated with woodland drop to only 1% and the fauna becomes dominated by species such as the 'garden chafer' *Phyllopertha horticola,* which is associated with the roots of plants in grassland, and two species of dung beetle (*Aphodius cf. sphacelatus* and *Onthophagus joannae).* Clover and vetch feeding species, such as the weevil *Sitona,* and others which feed on ribwort plantain, such as *Ceuthorhynchidius troglodytes, Mecinus pyraster* and *Gymnetron pascuorum*, also become relatively common. This led Robinson (2000a, 153) to suggest that by this time the area surrounding the river was predominantly a species-rich pasture. Elsewhere, the insect remains from the Middle Bronze Age Watering holes at Perry Oaks, Heathrow also are dominated by indicators for grassland and grazing animals (Lewis *et al.* 2006). Similar indicators for open areas were found at a number of Bronze Age sites examined as part of the survey by Elias (Elias *et al.* 2009). Recent work by the author on the insect faunas from a number of Middle and Late Bronze age ditches at the sites at Horton and the Imperial College Sports Ground near to Heathrow clearly suggest the same pattern of clearance at this time. Outside of the region, particularly in the Upper Thames Valley, the results from a number of archaeological sites clearly show that insect faunas of this type are very common from the early Iron Age onwards. These faunas are usually associated with pollen sequences that also suggest a cleared and farmed landscape (Robinson 1978; Robinson and Lambrick 1979; Robinson 1993).

This pattern in the pollen and insect profiles from greater London is also echoed in the archaeology. There is growing evidence for the presence of small-scale field systems in some areas of the Thames Valley from 2,000 cal BC onwards with the field systems at Hayes Common, Bermondsey and Heathrow appearing to be particularly early in the period (Merriman 1990; Drummond-Murray *et al.* 1997; Brown and Cotton 2000; Cotton 2004; Lewis *et al.* 2006). This is in contrast to some of the areas of Greater London outside of the Thames Valley itself that fail to develop such field systems until later periods (Poulton 2004). There is still an apparent lack of evidence for settlement itself in the Thames Valley during the Early Bronze Age (Merriman 1990; Brown and Cotton 2000; Cotton 2004). The Late Bronze Age and Early Iron Age (1000 – 700 cal BC) see the societies of the Lower Thames Valley undergoing a change in fortunes. Certainly by the later Bronze Age extensive field systems (some coaxial) with isolated settlements are found on the terrace gravels between the River Colne and Crane to the west of the modern city and in the area associated with the Ringwork at Queen Mary's Hospital at Carshalton on the River Wandle (Yates 2001).

Before the Middle Bronze Age there are few signs in the Middle and Lower Thames Valley for the clear display of power, status and wealth. There are a limited number of Neolithic and Early Bronze Age monuments, mainly in the area around Heathrow, but these seem to lack the presence of those seen elsewhere in the Neolithic and Early Bronze Age of Wessex and the Upper Thames Valley (Merriman 1990; 2000; Brown and Cotton 2000; Yates 2001). After this point, particularly in the Late Bronze Age and Early Iron Age, the Thames River Valley sees the large-scale deposition of impressive metalwork such as swords, spears and other weaponry, for example the Bermondsey shield, which dates to the end of the period. This deposition seems to be particularly near to areas associated with the larger field systems (Yates 2001). R. Thomas (1999), Merriman (1990; 2000) and Brown and Cotton (2000) all argue that this may relate to increased access to metal resulting from the establishment of trade routes along the Thames Valley for both raw ores and the metalwork that ultimately reach out towards the continent. However, it has been suggested that the drive to deposit this metalwork could also relate to rising water levels at this time and represent a 'coping mechanism' to deal with the social stress and change that resulted (Cotton 2004; Poulton 2004).

Certainly by 1000 – 700 cal BC the landscape of the gravel terraces of the Thames Valley seems to have been predominantly cleared (Branch and Green 2004). Many Late Bronze Age field systems and droveways continue in use or expand in many areas such as Runnymede, Downey, Lechlade, Wallingford, Heathrow and Marshall's Hill (Yates 1999; Merriman 2000; Crockett 2002; Lewis *et al.* 2006). Poulton (2004) suggests that although the overall number of settlements decreases in this period, they are far larger and more 'nucleated' than those seen in the later Bronze Age. In terms of settlement, therefore we could see a mix of larger farmsteads and larger circular enclosures, such as those at St. Mary's Hospital, Carshalton and near Heathrow, at this time. The swamplands upstream of what is now central London also seem to have been heavily utilised in this period with a range of river-front settlements identified, the best preserved of which is the complex at Runnymede-Petters (Needham 1991) which produced the insect faunas discussed above.

The Late Iron Age (400BC – 50AD), the period just before the arrival of the Romans, seems to have again been a fairly 'quiet' period of activity and settlement (Reece 2008; Sidell 2008). London seems to be located outside of the main Iron Age 'hillfort belt', though there are some examples of defended settlement at the periphery of the area (Poulton 2004; Wait and Cotton 2000). Merriman (1990; 2000) and Wait and Cotton (2000) suggests that there were small-scale agricultural settlements throughout the area but these were not to the same scale as those seen in the Late Bronze and Early Iron Ages and do not appear to have been part of planned landscapes. A number of pollen cores from outer London dating to this period confirm this to some extent, since they show regeneration in woodland at this period (Sidell 2008). Within the city itself recent pollen work at Poultry and Blossom's Inn confirm that a similar open landscape existed (Sidell 2008).

To the east of the Isle of Dogs, in the floodplain of the Thames basin, a different landscape type appears to have dominated. Here, an extensive freshwater 'peat' and alder carr seems to have developed during the Later Neolithic and Bronze Age (Devoy 1979; Meddens 1996; C. Thomas and Rackham 1996; Wilkinson *et al.* 2000). This was bounded by two periods of estuarine incursion (Rackham and Sidell 2000). Pollen analysis at several sites suggests the local vegetation of the area consisted of a mix of carr, fresh water marsh and estuarine conditions (Meddens and Sidell 1995; C. Thomas and Rackham 1996; Rackham and Sidell 2000). The depth and longevity of the formation of this deposit differs between sites suggesting a high degree of local variation (Wilkinson *et al.* 2000), but the area is mainly subject to estuarine conditions by the Iron Age. During the Bronze Age, this wetland landscape was traversed by a number of wooden trackways, the oldest of which being that from Neolithic Slivertown closely followed by that from Bronze Age Bramcote Green. The location of these sites is shown in Figure 3.1. The construction of the trackways ranges from rather simple constructs to quite complex structures. Those from the early phases at Tesco site are little more than brushwood bundles (Meddens 1996) and that from the later phases at Bramcote Green (Thomas and Rackham 1996) is nothing more than a series of logs laid one after the other. Contrasting against this is the complex construction of the trackway at Beckton (Meddens 1996) where a v-shaped cradle supported the track, or the four metre-wide pebble, burnt flint and silt 'causeway' that runs for 23 metres at the Hays site, Dagenham (Meddens and Sidell 1995; Meddens 1996). All of these trackways cluster in date between 1600 and 1000 cal BC. Brown and Cotton (2000) believe that this activity and the nature of some of the constructions suggest clear links to similar and contemporary structures in the Lee Valley, those at Runnymede and the wooden structures in the Essex Estuaries (Wilkinson and Murphy 1986, 1995).

Two insect faunas from these sites have been examined from the Bramcote Green trackways and the Hays site, Dagenham causeway. Those from the trackway at Bramcote Green were limited in terms of the insects recovered from the samples but broadly suggest a relatively open landscape dominated by pools of slow-flowing water, with a range of marsh plant species present. The Dagenham Hays causeway produced a far larger insect fauna, but again the assemblage was dominated by water beetle species associated with slow-flowing and stagnant water. Both the reed beetles and weevils present clearly suggest stands of rush and sedges. Large numbers of the small weevil *Tanysphyrus lemnae,* which is associated with duckweed, also were recovered. Specimens of two species of beetle that are now extinct in this country were recovered at Dagenham Hays: the ground beetle *Oodes gracilis* and the water beetle *Spercheus emarginatus.* Both species have probably declined due to the loss of wetland habitats, such as were crossed by the trackways in this part of London at this time.

Similar insect faunas and environments appear to have been associated with the Bronze Age trackways at Woolwich Manor Way, Newham (Elias *et al.* 2009), Bellot Street, Greenwich (Elias *et al.* 2009) and from the two trackways at Freemasons Road, Newham and Movers Lane, Barking, which I have just examined as part of the A13 redevelopment.

CONCLUSION

I hope that this chapter has demonstrated the extent to which archaeoentomology can contribute to a wide range of archaeological questions and issues. Although certain limitations in archaeoentomological data have been openly discussed here, frankly, no archaeological sub-discipline is wholly free of such methodological drawbacks. Archaeoentomology can be a powerful tool in archaeological reconstruction of ancient landscapes to niche habitats (e.g. fast-flowing, gravel-bottomed rivers, meadow or woodland). Finally, it should be obvious that there clearly is a paucity of archaeoentomological work carried out on prehistoric sites in the Lower Thames Valley, with some periods represented by only one or two sites. As a result, the sampling of waterlogged deposits of prehistoric date for insect remains should be a priority on archaeological sites in the Lower Thames Valley/ Greater London.

CHAPTER 4: USING INSECTS TO RECONSTRUCT THE 'VIRTUAL ENVIRONMENT': CLEARANCE IN ANCIENT WOODLAND

The previous chapter presented some examples of the strengths of archaeoentomology as a tool in the interpretation of archaeological deposits. In addition to being one of the few direct indicators of terrestrial climate, insect faunas produce a wealth of detailed information, both at the very local scale, directly 'on site', and across a wider area. This is not just landscape reconstruction; archaeoentomology also produces information on human land use and behaviour. However, it does have its limits. We can recover direct evidence for conditions in the past but often this leads to a very hazy reconstruction. It shows us potentially what the landscape was but not its actual, or if you prefer 'fine', detail. This is probably best explained by using some terms borrowed from a recent discussion by Chris Caseldine, Ralph Fyfe and Kari Hjelle when considering modelling of environments using pollen (Caseldine *et al.* 2008). They propose that the type of work seen in the previous two chapters is the 'virtualisation' of the landscape. By using our knowledge of the ecology of the species present, our understanding of how taxa are 'produced' and deposited and what happens to our material whilst it is in the 'archaeological record' we can attempt to reconstruct a 'virtual landscape'. It is our impression of the past, not necessarily its actuality. It only can be an interpretation. Unfortunately, this is not often what the archaeologist, or indeed the reader of this book really wants. The desirable 'end product' is the 'truth'. Caseldine *et al.* (2008) refer to this as an attempt to produce a 'visualisation' of the landscape. To put it bluntly, this is the attempt to say what the landscape actually looked like and allow a picture to be drawn for display, rather than suggesting a notional and mental reconstruction. Caseldine *et al.* (2008) see this as a move from suggesting 'plausible' landscapes to 'faithful landscapes'.

What follows should not be seen as an attempt to undermine the use of archaeoentomological data to reconstruct ancient environments, but rather an attempt to recognise openly its limitations. It examines a particular archaeological issue and the suitability of archaeoentomological data to address it. In this case are we asking too much? We know we can use insects to show that a range of different components could plausibly make up the landscape (for example fresh water swamps, alder woodland, lime dominated terrace woodland, meadowland), but can we use this information to faithfully reconstruct the relative proportions of these elements in the landscape?

To be fair I don't think that we archaeoentomologists, who are keenly aware of the problems of interpretation (for example those outlined by Kenward 1975b, 1978), would ever attempt such an exercise. Perhaps the only exception was early work by Robinson (1983) that dealt specifically with the ratio between pasture and arable land.

Where this issue has become important recently is the way in which insect remains have been used to deal with one particular aspect of early woodland in northern Europe; specifically, the degree to which early Holocene woodland contained large clearings or glades (Kenward 2006; Whitehouse and Smith 2010; Smith *et al.* 2010).

The 'traditional' view of the primeval forest, primarily based on palynological evidence, is that in the Mesolithic and the Neolithic there was a dank, dark and continuous blanket of trees from eastern Europe to the coast of Ireland. This is the 'climax', primeval closed canopy or 'high forest' model (Iversen 1960; Godwin 1975; Peterken 1996). Where openings in the forest are present, they result from the fall of dead trees, actions of beavers or erosion.

In 2000, Franz Vera published a book that fundamentally questioned this assumption. Vera held that a large proportion of the landscape might have consisted of clearings and glades. Vera noted that oak and hazel occurred in relatively large proportions in the pollen diagrams of Northern Europe. From his experience as a woodland conservation officer, Vera knew that these tree species are light demanding and favour the woodland edge. In addition, he knew that these species are declining in most woodland preserves in northern Europe at the present time, as 'full canopy woodland' is re-established due to conservation-based woodland management. Vera suggested that there must be a missing factor which explains why oak and hazel survived in the early Holocene forest, but has been out-competed by shade tolerant tree species today.

He suggested that large-scale clearings were present (providing the oak and hazel with a suitable habitat to thrive) and that these were opened up and maintained by large grazing mammals such as aurochsen, bison and wild horse. Skeletons of these animals are found in deposits from this age. Vera suggested, therefore, that not only have the pollen diagrams been 'misread', but also that the present model of the appearance of ancient woodland needs to be reconsidered.

This is not a minor point. It has clear implications for the nature of modern woodland conservation. Perhaps, rather than returning commercial woods to closed canopy woodland, in an attempt to return them to 'primeval forest', cattle should be run in woodlands to help maintain open areas. However, it would suggest that many of us working in palaeoenvironmental reconstruction of past landscapes have essentially got it wrong for many years. For example how would such open woodland fit into the landscape picture outlined above for the Greater Thames Valley in this period? For archaeologists it also changes our view of how Mesolithic and early Neolithic people moved through the landscape and how they used it.

There is at present a very heated debate occurring between two camps of scientists that support or refute Vera's suggestion. The disagreement seems to occur between archaeologists and conservation officers of several different flavours who favour Vera's hypothesis and palynologists who do not. Of course the former accuse the latter of protectionism.

Where, unfortunately, the insect remains have entered this debate is that both camps have tried to use aspects of the faunas recovered from this period to support/ reject their argument. For example, do the insect remains from Mesolithic and Neolithic forests support the suggestion that oak and hazel are important tree species? If this were the case then clearly Vera's argument would have to be correct to some extent. Figure 4.1. (Whitehouse and Smith 2010) presents the number of species of beetles that are associated with a range of trees from 17 different archaeological sites from this period. It clearly can be seen that many species from these sites are frequently associated with oak or beech, and to a lesser extent with hazel. Does this data support Vera's hypothesis?

The short answer is no. If you look at Figure 4.2, you can see this is not the whole case. The number of modern insect species associated with each tree species is presented and, as you can see, there are many species that live on oak and pine. In fact all we are seeing in the archaeological record is nothing more than a reflection of this statistic (i.e. that oak and pine support a wide range of beetles). A clear example of this problem can be seen when we consider the occurrence of insect species associated with alder in the archaeological record. Many of the sites in the Thames Valley were clearly situated within alder swamps on the basis of the pollen and plant macrofossil evidence (Robinson 1991, 2000a; C. Thomas and Rackham 1996; Scaife 2000). However, this is rarely reflected in the insects present; there are very few beetles recovered from Mesolithic/ Neolithic sites that are associated with alder. The explanation for this apparent contradiction between pollen and plant macrofossil evidence and the insects can be explained easily. There are actually very few species of beetles that are associated specifically with alder, in fact only 11 in this country (Smith and Whitehouse 2005). In addition, these species are relatively uncommon at present and probably this was also the case in the past. The fact that these species are thought to require very old, well-established alder carr to be successful may also help explain this trend in the data. As a result it seems highly likely that alder is under-represented in the archaeoentomological record (Girling

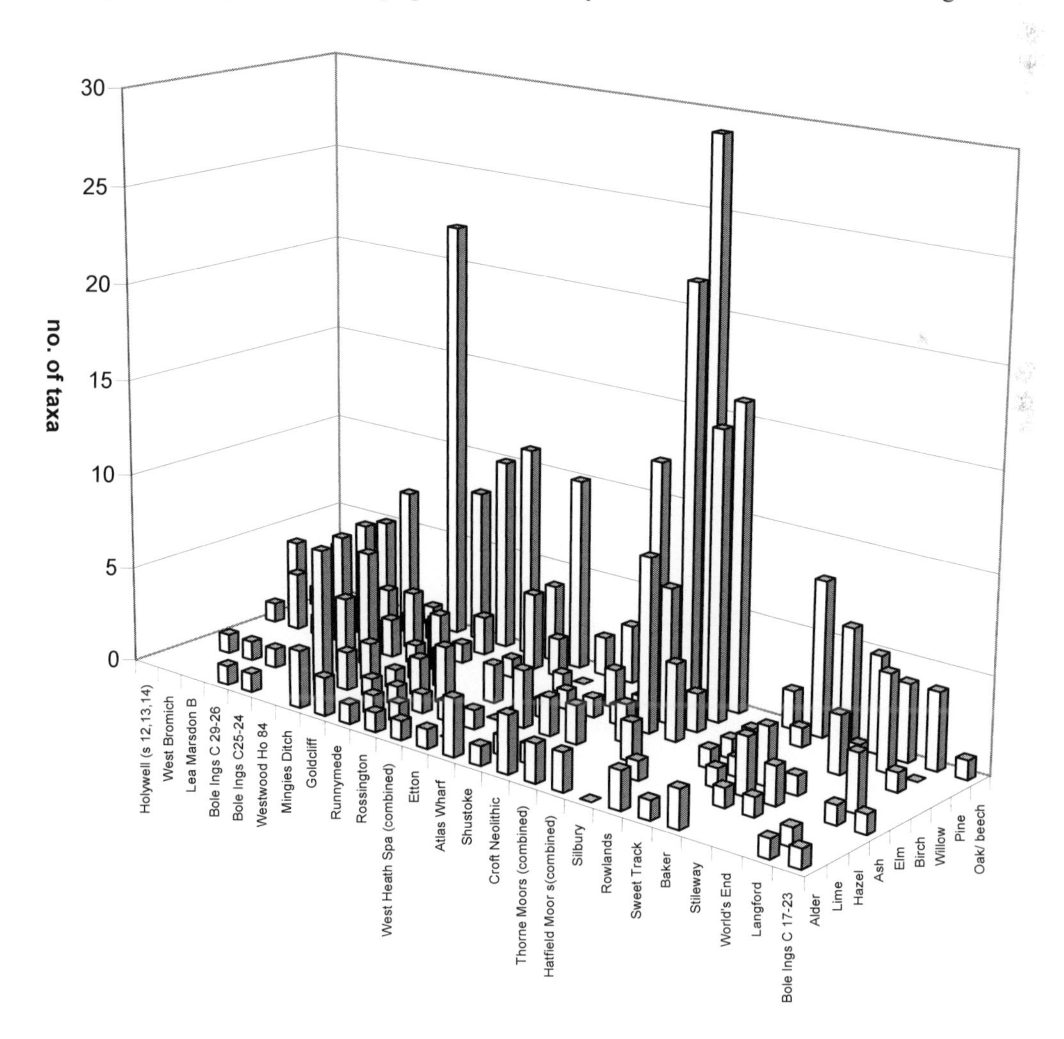

Figure 4.1. The number of species of beetles that are associated with a range of trees from 17 different archaeological sites from the Mesolithic and Neolithic

1985; Robinson 1993; Smith *et al.* 2000; Smith and Whitehouse 2005).

Here we have the major problem with the use of the 'mosaic approach' in archaeoentomology; a method outlined in Chapter 1 and used in Chapters 2 and 3. We have evidence for a certain type of ecosystem (a 'virtual landscape') but an inability to determine the significance of that component in the landscape (so to borrow again from Caseldine *et al.* (2008) an inability to 'visualise'). For example, and to be intentionally fatuous, there is no way to relate the number of dung beetles in a deposit directly to the number of animals present in the past.

This leads us on to another example of how insect remains have been used to support Vera's argument. Figure 3.2. also showed the proportion of beetles associated with the dung of large herbivores and open ground from the Mesolithic and Neolithic sites. This can run as high as 10%. Vera's proponents would dearly like to use this data to suggest that there are large herds of aurochsen, bison and other larger sized herbivores during the Mesolithic and that this supports the argument for large clearings. The possibility of persistent and large clearings may also shed light on the circumstances that lead to the deposition of 60,000 stone tools during the Mesolithic at West Heath Spa, a location well away from the usual areas of the floodplain, often used by grazing or watering animals during this period (Sidell *pers. com.*). Certainly, both Girling (1989a) at West Heath Spa and Robinson (1991) at Runnymede Bridge try to argue that the presence of dung beetles and other indicators of clearance in the insect fauna supports the argument for early Neolithic forest herding or for the adoption of tending domesticated cattle as a substation for the hunting of wild ungluates (J. Thomas 2004). However, exactly how many animals this 10% dung beetle fauna really represents is open to question? The honest answer, of course, is that we have no clue.

The next time you are in the countryside take a bucket of water with you and throw in a cowpat (this is very entertaining game to play with small children). The hundreds of beetles that can emerge from even a relatively small pat will amaze you. Now imagine if that pat, which only represents several seconds of activity by a single animal, finds its way into one of my archaeological samples. It could lead to the impression that a lot of dung and therefore lots of animals were present. There is also the rather indelicate problem of just whose pat is it? It could belong to a bison or a domesticated cow and, therefore, would indicate clearance. Or it could be a creature of dense wood, like red deer, which would not suggest clearance. Unfortunately, it is often difficult to precisely link a specific species of dung beetle with the dung of one particular species of large herbivore alone. As a result, archaeoentomologists with assemblages including dung beetles can rarely say for certain if they represent wild herbivore or domesticated livestock grazing. However, there may be some light at the end of the tunnel. A recent modern analogue study undertaken at both Dunham Massey, Cheshire and at Epping Forest, Greater London has indicated that that proportions of dung beetles in insect faunas of 10% or below probably do not represent particularly large concentrations of grazing animals (Smith *et al.* 2010).

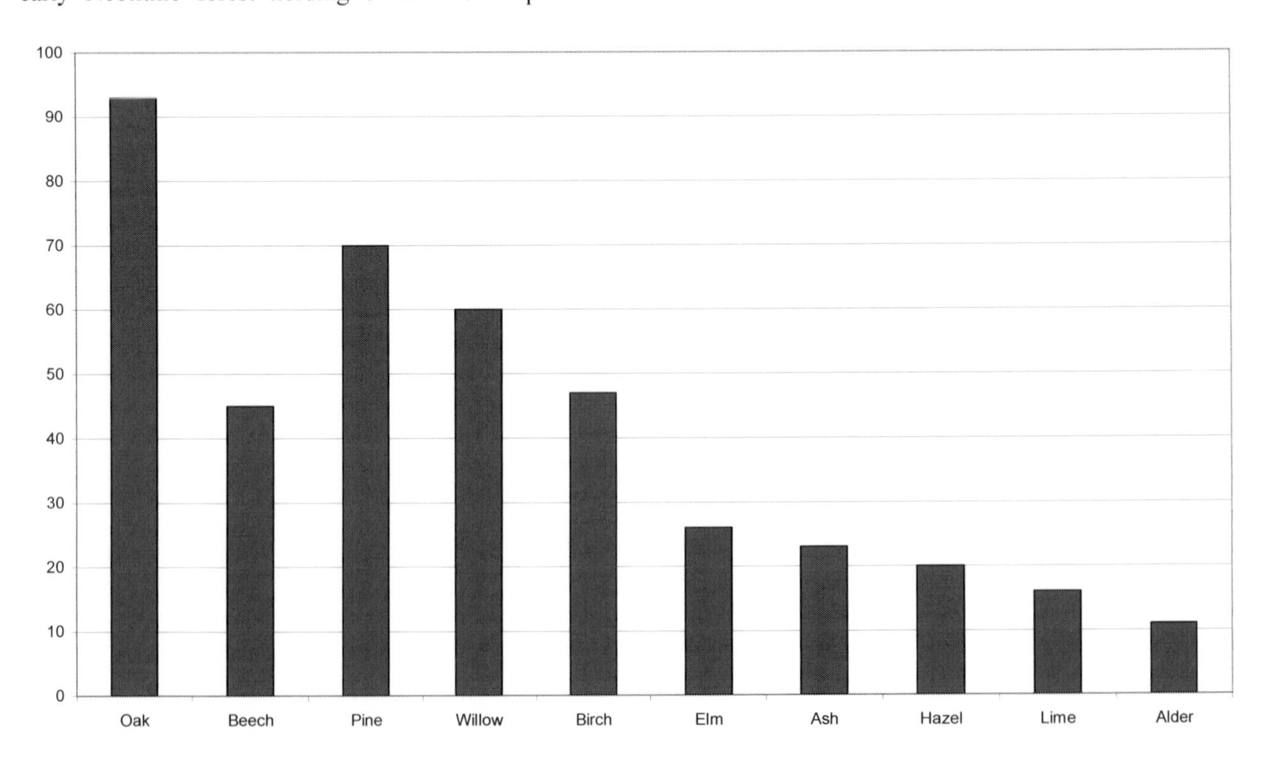

Figure 4.2. The number of modern species of beetle associated with a range of common trees (Smith and Whitehouse 2005)

The problem here is that we are increasingly attempting to extract more detail from our data. Often I feel that the questions we are beginning to ask are too 'fine grained' for the data we have. They are more appropriate for landscape reconstruction on the level of the Caseldine *et al.* (2008) 'visualisation' rather than as 'virtualisation'. This does not mean we should not ask these questions or develop techniques to better address them; but we do need to accept that often we will only be able to give 'plausible' rather than 'faithful' reconstructions of past landscapes.

CHAPTER 5: ROMAN LONDON: ROMAN INSECTS?

This chapter discusses the nature of Roman London and its insect faunas from shortly after the Claudian invasion in AD 43, during the time of London's earliest Roman settlement through to the end of the Roman period in the 4th century. In contrast to the previous chapters, where essentially 'natural landscapes' were described, this chapter will concentrate on an urban landscape. The analyses in Chapters 2 and 3 worked at the large and medium scale, in other words regional and local landscapes. In this chapter the results presented involve extremely local environments (often related to habitation) and the process of identifying the nature of individual archaeological contexts. This leads to directly suggesting how human agency shaped and formed these deposits. In terms of pure archaeology this is often the level at which some of the most striking and important results are achieved.

In this chapter we will discuss two such results:

1) From the earliest phases of urban occupation at Roman London the insects present are predominantly urban in their nature; what does this tell us about living conditions and waste disposal at this time?
2) There is a high incidence of 'granary pests' in all deposits; what does this tell us about Roman grain supplies and production?

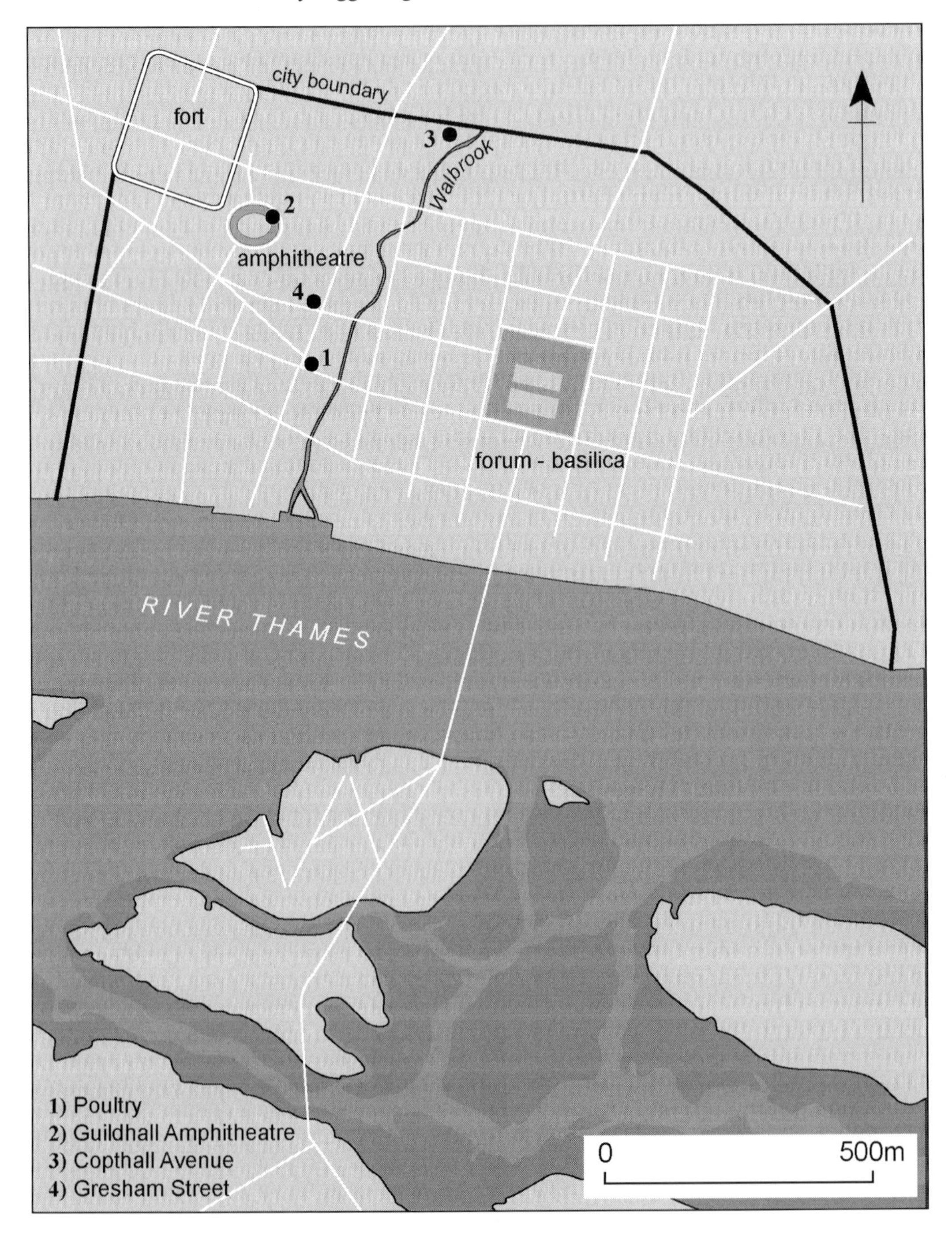

Figure 5.1. Map showing location of the Roman sites mentioned

In terms of archaeoentomological interpretation two issues will be raised. We will see the start of the discussion of urban archaeoentomology in general and the role of the use of specific 'indicator species' or 'packages' in particular. The location of the various sites discussed below is illustrated in Figure 5.1

BACKGROUND TO EARLY ROMAN LONDON AND THE INSECT FAUNAS RECOVERED

When I worked on the insect remains from the site at Poultry in Central London I completely failed to realise their significance. Subsequently, I have come to realise that this site, and the material from it, represents the earliest evidence we have for both the foundation and early occupation of Roman London. Some idea of the how dramatically this site and its archaeology have changed our perception of Early Roman London can be seen by examining the state of our knowledge of this time before the Poultry excavation occurred. This was outlined in a number of texts written in the 1980s and 1990s (Marsden 1980; Merrifield 1983; Hall and Merrifield 1986; Perring 1991, Perring *et al.* 1991; Milne 1995) and in a more recent summary of the situation by Hassall (2000). The actual date of the foundation of the settlement of Londinium apparently remained a mystery. Certainly it was clear from the Roman texts, mainly that of Tacitus and Dio, that by 60 AD London was a relatively rich and well-off town of traders presumably centred on the legionary fort and the river crossing at Southwark (Perring and Brigham 2000; Mattingly 2006) and the bankside docks (Rowsome 2008). However, the time of the founding and nature of the early settlement was not clear. Before the 1980s, much of the available evidence for the date of founding of London came from excavations across the river at Southwark (Merrifield 1983; Perring 1991; Milne 1995; Perring and Brigham 2000; Sheldon 2000; Drummond-Murray *et al.* 2002; Cowie 2003; Mattingly 2006) rather than from central London. In Southwark, a number of packed gravel roads had been found that seemed to lead towards a point just south of the river, presumably the location of the bridge across the Thames. Much of the dating evidence, mainly coins, suggested that most activity occurred between 50–60 AD, which implied a seven year gap between the Claudian invasion in 43 AD and the foundation of the settlement at London.

Across the river in central London, evidence was even patchier and mainly concentrated on a range of timber-framed buildings in Lombard Street and Gracechruch Street preserved by charring during the Boudiccan revolt (Merrifield 1983; Perring 1991; Milne 1995). For Merrifield (1983, 43) and Marsden (1987), this suggested a regular layout of timber buildings and gravelled areas that might be associated with the Roman Army, though this has been questioned (Millett 1990; Perring 1991; Milne 1995; Dunwoodie 2004). Certainly, early development in the area around Cornhill seems to initially fit the regular grid of streets favoured by the Romans (Rowsome 2008). Other areas of the settlement, such as the area to the west of the crossing of the Walbrook seem to have developed in a more organic way (Rowsome 2008). In addition, at sites such as Newgate and Cheapgate the remains of round houses built, as Milne (1995, 45) rather quaintly puts it, in the 'old native style' were found (Perring *et al.* 1991). This, at least, suggested some mix of settlement styles and population in the initial phase of settlement (Perring *et al.* 1991; Perring and Brigham 2000). To the northeast of this area at Aldgate, excavations in the 1970s (Chapman and Johnson 1973) revealed a short length of deep ditch with an 'ankle break' at the bottom. This form of feature is normally associated with the Roman military and led Merrifield (1983) to suggest that this is the possible location of the first fort (Mattingly 2006). Once again this has been questioned. Later discoveries of similar ditches at Fenchurch Street and Rangoon Street seem to be associated either with roads or small enclosures (Perring 1991; Milne 1995; Perring and Brigham 2000). Similar ditches from this period found at the Baltic House excavation have been interpreted as parts of defensive boundaries (Howe 2002; Perring and Brigham 2000). The situation in the area to the west of the channel of the Walbrook was, in the middle 1990s, no clearer. Though several excavations showed that the area was fully built up by the 2nd century AD, only one site, Newgate Street, contained evidence of early occupation from before the Boudiccan revolt of AD 60. It was presumed that this was an industrial area and, initially, probably peripheral to the settlement.

This scatter of hints and scant finds lead Merrifield to suggest that:

'Somewhere within them [the small maze of streets in the area] *we need a site or sites with opportunity for a fully scientific investigation'* (Merrifield 1983, 34).

The excavation at the Poultry site made this possible (Rowsome 2000, Hill and Rowsome 2012). It is worth noting that, before Poultry was excavated, none of the early Roman sites previously excavated in London were waterlogged. This probably means that there was no potential for insect analysis, or many other types of archaeoenvironmental analyses, to have occurred anyway. However, despite this, it is clear that environmental sampling was extremely limited at some of these sites, leaving some doubt as to whether a full environmental sampling programme was undertaken.

Of course the other question commonly asked about early Roman London is – was it primarily a military or a civilian settlement? This is an argument that continues to this day (Perring 1991; Millett 1994, 1996; Milne 1995; Hassall 2000: Yule 2005), with Mattingly (2006) suggesting that it was probably a combination of both.

THE SITE AT POULTRY

In 1995, during the construction of the new building at Poultry, the Museum of London Archaeology Service undertook a detailed excavation. In addition to later Saxon and Medieval deposits (which will be discussed in Chapters 7 and 9) they encountered a range of roads and civil buildings dating to the earliest Roman phases of London's past (Rowsome 2000, Hill and Rowsome

2012). The layout of this part of the pre-Boudiccan town is show in Figure 5.2.

Preservation on site was very good with many deposits waterlogged and containing substantial timbers and large quantities of other biological evidence. The earliest Roman feature was the remains of the gravel packed main road, the *Via Decumana*, which continues east, crosses the Walbrook and forms a T-junction with the road coming north over the river from Southwark near Cornhill. A timber drain from beneath the first phase of road building at Poultry gave a dendrochronology date of 47 AD (Rowsome 2008; Tyers 2008). This is the earliest date for the foundation of Roman London and in itself suggests substantial activity before AD 50. The area saw the rapid development of a range of rectilinear earth and timber buildings backed by yards full of sheds and outhouses. These yards also contained raised post-built structures that were probably grain stores (Rowsome 2000, Hill and Rowsome 2012). The impression, therefore, is of an area of mixed commercial and residential properties. The development of this form of settlement at Poultry presumably echoes events across the Walbrook river at this time (Rowsome 2000).

DESCRIPTION OF THE INSECT FAUNAS FROM THE PRE-BOUDICCAN POULTRY SITE

Eight insect faunas come from deposits dating to before 60 AD at the Poultry site (Smith 2012 – The ecological summaries for these samples are shown in Figure 4.3). Three were associated with 'Structure 4', which backed onto the *Via Decumana,* and the remainder with a series of deposits from the open yards at the back of the building plots. I will outline the nature of the insect faunas recovered from these deposits in detail since this is the first occasion on which we have met 'urban faunas' such as this in this book. This detailed description also helps to set the tone for the chapters that follow where these types of insects, and the 'groupings' they fall into, can be introduced, since they are also commonly recovered in later periods from London.

Insect indicators for settlement, housing and living conditions at Early Roman London

The insect faunas recovered from the earliest Roman deposits at Poultry, especially those from Structure 4, are typical of dense urban settlement in the archaeological record (e.g. Hall and Kenward 1990; Kenward and Hall 1995; Hellqvist and Lemdahl 1996). Since his initial

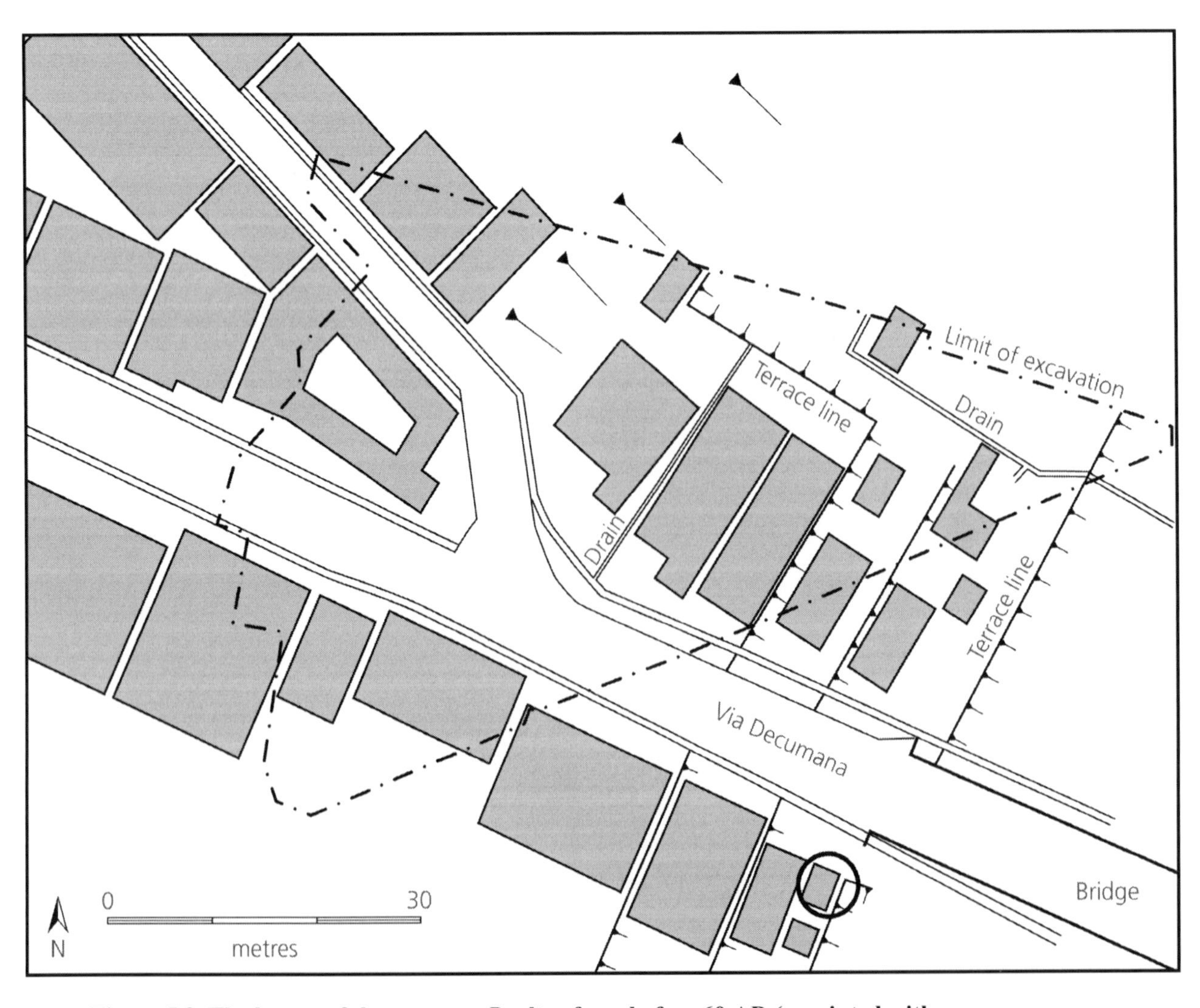

Figure 5.2. The layout of the streets at Poultry from before 60 AD (reprinted with permission from Rowsome 2000)

work in the 1970s on Anglo-Scandinavian material from the Coppergate site at York, Kenward has established that there is a common set of insects associated with urban archaeological settlement and waste (for example Buckland *et al.* 1974; Hall *et al.* 1983). One interesting aspect of this 'group' of insects is that it is ecologically quite discordant. The 'groups' are not quite what we would expect as modern entomologists. There is a strange blend of species.

Of course there are species: such as the 'woodworm' (*Anobium punctatum*), the 'death watch beetle' (*Xestobium rufovillosum*) and the 'powder post beetle' (*Lyctus bruneus*), that are still common inhabitants of the urban landscape. This is, to some extent, also true of 'spider' (Ptinidae) and 'plaster' (Lathridiidae or Cryptophagidae) beetles. Spider beetles were common, but often-unnoticed, inhabitants of cold, damp and ill-kept housing up to the recent past. Today, this habitat is only found in the kind of establishment where the central heating fails to come on and the vacuuming is never done well. Student accommodation immediately springs to mind as our nearest modern equivalent. This can be clearly illustrated with a real-life example. A few years ago, a group of students presented me with the fat from a grill pan that was infested with spider beetles. It was the usual tale of young angst and woe. They had cooked bacon three weeks earlier and not cleaned out the grill pan. Plaster beetles also occur in old and cold housing where they feed on mould. If there is a small beetle scampering around in the bread bin it is probably one of these. A few lines back I did say 'in the recent past'. Your grandparents, or their parents, would have commonly seen these species in their housing along with other urban denizens such as 'silverfish' and 'fire brats'. For the uninitiated, the range of creatures regularly infesting housing once was common enough to deserve their own 'natural history book'. Ordish's (1960) excellent book '*The Living House*', more recently updated by Mourier *et al.* (1977), catalogues the diversity of 'wildlife' commonly infesting 'traditional' housing in Britain. It is the advent of central heating and the vacuum cleaner that has sadly resulted in the demise of much of the natural habitat of our traditional house insects (or pests as some might describe them).

A number of species that were very common in the deposits at Poultry and York, and from other urban archaeological sites, do not occur in urban environments

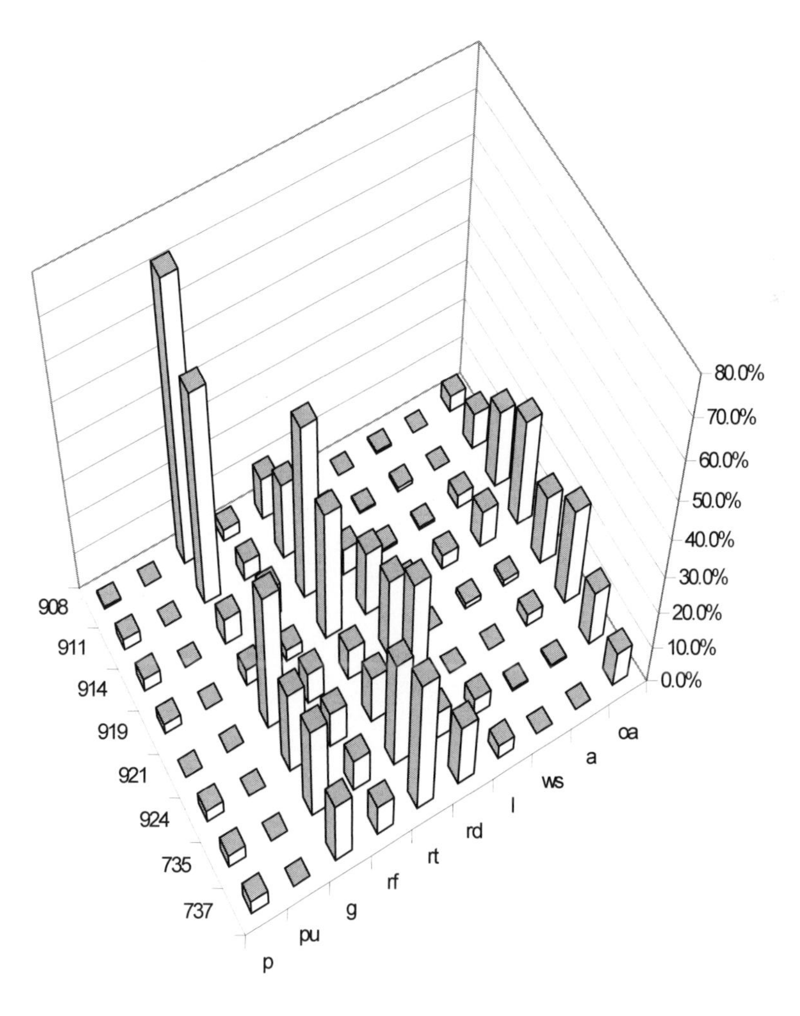

Figure 5.3. The ecological summary statistics for the insects from the pre-Boudiccan deposits at Poultry (Codes are defined in Table 1.2)

today. A prime example of this is the rather strange, small and blind colydiid beetle *Aglenus brunneus (*Kenward 1975a, 1976). It frequently occurs in most urban archaeological deposits, sometimes in very large numbers. It also is commonly associated with many of the insect pests of housing mentioned above, and clearly is not associated with farm wastes or agricultural areas. Today, it occasionally is encountered in dark cellars, rotting posts, mushroom beds and dead wood that have been incorporated into wet mud. Although not rare, it is not a particularly common species, and I certainly have never seen a live specimen.

Kenward has labelled this group of species the 'house fauna' (Hall and Kenward 1990, Kenward and Hall 1995). This is, of course, a slightly misleading identification because this group is not solely limited to archaeological housing. However, they are certainly typical of urban archaeological settlement. You can get an idea of what species form this group by looking at the species lists in Appendix 1, where members of the 'house fauna' are indicated by an 'h' for their habitat code. They are also outlined in Figure 1.2 in Chapter 1. Carrott and Kenward (2001) have now added several taxa to this 'house fauna' to make 'supergroup A'. This now includes the various 'woodworms' mentioned above, several 'darkling' or 'hide' beetles such as *Blaps, Tenebrio* and *Demestes* species. In addition, several common ectoparasites of both humans and animals are now included; such as the human flea (*Pulex irritans)*,and lice (i.e. *Damalina* spp.).

Many urban and settlement deposits also contain a range of species that really are only associated with wet, detritus-filled mud today; the type of stuff that releases a smell like a wet dog when you put your foot in it. Typically these include a range of small 'rove' beetles such as the *Oxytelus* and *Trogophloeus* species. These species were once included in Kenward's various 'r' groups spread across different types of rotting rubbish (Kenward and Hall 1995). Now they have been grouped along with various dung beetles to form 'supergroup B' which is held to be typical of gullies and wet floors and other external areas in the archaeological record (Carrott and Kenward 2001), though they occur in some house floor deposits at Anglo-Scandinavian York too (Kenward *pers. com.*).

What on earth are species associated with dry hay, rotting wood, forest floors and lakeside mud doing together in dense urban settlement in the past? Kenward (Hall and Kenward 1990, Kenward and Hall 1995) holds that the key factor, in addition to no central heating and vacuum cleaners, may be the presence of earth or 'beaten' floors in the housing. Wet and muddy areas of earthen floors with a covering of litter would provide suitable homes for many of these species. Frequent use of wattle and daub walls might also provide a haven for these and many other species of beetle.

Often, to the modern entomologist, groups such as this can appear to be oddly sorted and discordant. However they do make coherent archaeological sense. These groupings also have been supported and improved by sound statistical work (i.e. Carrott and Kenward 2001; Kenward and Carrott 2006). We will return to this issue in Chapter 11 when we look at the statistical comparison of the faunas from London.

Although we can recognize ecological groups of taxa such as the 'house fauna' in the archaeological record, what these groups actually mean (i.e. in terms of specific living conditions) is not completely understood. Do they result from the mixing of different materials and their accompanying insect faunas in the archaeological record or are they actually functioning biological communities? If the latter is the case, it does have some rather uncomfortable implications for community biology and suggests that relationships between species can be rather flexible.

Indicators for filth, rubbish and decay

Back to early Roman London; in addition to the 'house fauna' and the species associated with 'puddled mud' most of the samples from the early levels at Poultry contain insects that occur in a range of decaying plant materials and settlement rubbish, in essence a blend similar to the inside of compost heaps. Kenward (Hall and Kenward 1990; Kenward and Hall 1995; Carrott and Kenward 2001) describes this group of species as the 'decomposer community' and they make up the bulk of Kenward's 'r' groupings and Groups C and D in his revised system (Carrott and Kenward 2001). Taxa typical of this group include the histerid 'pill beetles', Ptilidae 'feather wings', the majority of the staphylinid 'rove' beetles and species of *Anthicus*. The dominance of this group is reflected in Figure 5 where the 'rt' (rotting material) grouping accounts for 25–45% of the total fauna recovered. The presence of decaying urban detritus and rubbish is also indicated by species such as the small *Cercyon* beetles. These are the small black and yellow round beetles that live in the wet material, such as that found at the top of your compost heap.

Potentially some the *Aphodius* 'dung beetles' probably also inhabit this type of material. Traditionally, finds of these species in archaeological settlements were explained by suggesting that they either came in as inhabitants of dried cow pats used as fuel (i.e. Buckland *et al.* 1974) or as 'accidental tourists' from the wider environment (Kenward 1975b, 1978). Kenward (Kenward and Hall 1995) has recently suggested that these species probably had a more direct involvement with urban deposits. In particular he suggests that *Aphodius granarius*, *A. lapponum* and *A. fimetarius* can live and breed in accumulations of wet decaying settlement waste. Certainly these three species are common at Poultry and logic would suggest that Kenward is probably right. The alternative is to suggest that large numbers of cattle were present in the yards.

Whatever the pattern suggested by the beetles, the fly pupae recovered in the early Roman deposits at Poultry

(Smith 2012) portray an obvious and rather squalid story. Many deposits contain numbers of the common housefly (*Musca domestica*), its close relative *Hydrotaea dentipes* and the stable fly (*Stomoxys calcitrans*). These are all common inhabitants of decaying rubbish and waste in human settlement. This is particularly true where rather liquid rubbish, excrement and carrion has been scattered (K.G.V. Smith 1973, 1989). Similarly, the *Sepsis* fly is quite common in the material sampled at Poultry and appears to have a fairly consistent affinity for cess (K.G.V. Smith 1973, 1989). Another material that seems to have been 'blended' with this mess left lying around is stabling waste. This is clearly suggested by the two abdomens of *Damalina* biting lice found in one of the external dump deposits from the yards (Smith 2012).

Lice can also be very helpful in archaeological interpretation since they are often associated with only one individual species of domestic animal (how successful this can be is seen in the example of the Medieval pond deposits from St. Mary Spital (Smith and Chandler 2004) that will be discussed in Chapter 9. However, here at Poultry we could not pull off this trick since the heads of the lice were missing and as a result they, and their potential hosts, could not be identified to species level.

So it seems that the insects from the pre-Boudiccan deposits at Poultry suggest that timber framed buildings with earthen floors were present. It appears that both internal and external surfaces were scattered with accumulations of decaying plant material, food waste and cess. This is a fairly repellent view and rather upsets the 'niceness' of some of our reconstructions and ideas of the past. This has lead to some speculation about these piles of *ordure* functioning as 'central heating' and releasing heat through biological decay into housing (Coope 1981). Earlier in Chapter 1, I outlined some problems with looking at this phenomenon in such a straightforward way and that the life of deposits after they are abandoned or buried needs to be considered.

Life outside of the buildings

Compared to the samples associated with Structure 4, many of the faunas from the early Roman external yards at Poultry appear to contain a higher proportion of beetle species that come from a natural 'outdoors' environment, rather than man made environments (Smith 2012). Most are a range of Carabidae ground beetles that are usually associated with areas of scuffed ground. Also present are a range of weevils that suggest the types of plants that grew in the yards. For example, clover (*Trifolium* spp. - the food plant of the *Sitona* and *Hypera* species), docks (*Rumex* spp. - the food plant of the *Rhinocus* species), cranes-bill (*Geranium* spp. - the food plant of *Zacladus affinus*) and plantains (*Plantago* spp. - the food plant of *Mecinus pyraster* and the *Gymnetron* species) are all indicated. Of course, there is another possible explanation: the presence of these species has often been used to infer the presence of stabling waste in a deposit (Kenward and Hall 1997; Smith and Chandler 1994). However, in this case, many of the other species of beetle which appear to have a specific association with decayed stabling material are absent from these samples, or only occur in low numbers. This probably suggests that this 'outdoor fauna' is derived from the surrounding yards and open areas, or as background fauna, rather than through the direct incorporation of field hay.

Perhaps more difficult to sort out is the origin of a range of water beetles recovered from the deposits at Poultry. The *Hydroporus* and *Bagous* species are all typical of slow-flowing water. The weevils *Notaris acridulus* and *Leiosoma deflexum* are usually associated with waterside vegetation, in this case reed sweet grass (*Glyceria maxima* (Hartm.) Holmb.) and waterside buttercups such as marsh marigold (*Caltha palustris* L.). Whether dredging from a nearby body of water has incorporated these species into the deposit or they result from the disposal of material accidentally contained with collected reeds and sedges is not clear.

Human health and hygiene

It is traditional at some point in the preparation of a consultancy report on urban insect faunas to discuss the implications of insects for human health. In particular two species of flies from Poultry, the housefly (*Musca domestica*), and the stable fly (*Stomoxys calcitrans*), are seen as potential vectors for a range of detrimental pathogens such as salmonella, typhoid, diarrhoeal infections and possibly the transmission of poliomyelitis (Oldroyd 1964; K.G.V. Smith 1973). Indeed the housefly is also known as the 'dark destroyer'. Kenward and Large (Kenward and Hall 1995: 762) have suggested that other insect species breeding in scattered rubbish in settlements could potentially carry pathogens and internal parasite eggs, such as those of the *Trichuris* worms, into human housing. Indeed at Roman and Anglo-Scandinavian York large quantities of the eggs of intestinal worms have been recovered (A.K.G. Jones 1983, 1985; Kenward and Hall 1995). Other species of parasite, in this case external, have also been recovered. The single abdomen of a female human flea (*Pulex irritans*) from sample 675 suggests that the inhabitants of this part of Roman London probably carried the usual infestations of ectoparasites. I say usual, because Kenward's work in Roman and Anglo-Scandinavian York (Hall and Kenward 1990, Kenward and Hall 1995) suggest that people carried considerable populations of both fleas and human lice (*Pediculus humanus*). Once again these parasites have the potential to act as vectors for disease.

All in all, the above paragraph could lead to the conclusion that the inhabitants of Early Roman London were waiting to 'drop like flies' from a range of diseases carried by ...flies. However, this is not necessarily the case. In Chapter 8 I will outline why I believe that this aspect of potential vectors as indicators for disease could be exaggerated.

Grain pests

One aspect of the insect faunas from Poultry is particularly striking. Pests of stored products in samples 908 and 911, from Structure 4, account for 73% and 56% of all the individuals present respectively (Smith 2012). Pests of stored grain also account for 20% – 40% of the insects in the material from the open yards (Smith 2012). *Sitophilus granarius,* or the granary weevil, is most frequently encountered. The granary weevil is a common pest in granaries where both the larvae and the adults feed on whole grain which is in the early stages of spoilage (Coombs and Woodroffe 1963; Hunter *et al.* 1973). It is a species that can be very destructive and can cause a considerable loss of stored grain if an infestation is allowed to get out of control. Certainly, such high proportions demonstrate that this area of town was heavily involved in grain storage, processing and usage.

Other pests of stored products present at Poultry attack grain that has been broken and that has become wet and mouldy. Often this is material that has already been infested and damaged by *S. granarius.* This second group of taxa usually are described as 'secondary' species in the natural succession of infestation of stored grain (Coombs and Freeman 1956; Hunter *et al.* 1973; Freeman 1980). These are species such as *Oryzaephilus surinamensis* 'the saw toothed grain beetle', *Laemophloeus ferrugineus* 'the flat grain beetle', *Palorus ratzburgi* 'the small-eyed flour beetle' and *Tribolium castaneum* 'the rust red grain beetle'. However, they will also attack a range of other stored products, and in the case of the last two species this can include flour and bran meal (Salmond 1957; Hunter *et al.* 1973; Freeman 1980). Their recovery might provide limited evidence for the presence of flour and milling on site. *T. castaneum* is a species which is not believed to overwinter successfully in unheated warehouses and today is often considered to be a pest of imported grain (Solomon and Adamson 1956). Although it is pure supposition, its presence might possibly suggest evidence of imported rather than 'home grown' grain. There is some support for this argument. Charred grain from the Fenchurch Street excavation contained non-native einkorn, lentils and bitter vetch and is thought to be Mediterranean in origin (Milne 1995; Perring and Brigham 2000; Dunwoodie 2004).

Several other species such as *Tenebrio obscurus and Alphitobius diaperinus* are associated with the last stages of the breakdown of grain as it rots and moulders in sheltered stores. Similarly many of the species of Cryptophagidae, Lathridiidae and Ptinidae seen at Poultry are inhabitants of waste left in grain stores (Coombs and Freeman 1956; Coombs and Woodroffe 1963; Hunter *et al.* 1973).

In passing it is worth noting that in 2007 similarly aged deposits from Gresham Street just to the north of Poultry were investigated. The four insect faunas recovered were essentially similar to those seen here. Notably again there is a clear dominance of grain pests recovered. The implications of insect faunas dominated by grain pests, the origins of these species and what their presence implies for Roman grain production will be discussed further at the end of this chapter.

ROMAN LONDON AFTER THE REVOLT: AD 60 – 200

The later part of the first century AD, and the whole of second century AD, is a period that is well represented in terms of insect faunas. The history and nature of the settlement of London at this particular time, therefore, warrants some discussion. Certainly the archaeological record is much better explored than previous periods, resulting in a relatively detailed impression of London at this time (Marsden 1980; Merrifield 1983; Hall and Merrifield 1986; Perring 1991; Perring *et al.* 1991; Milne 1995; Mattingly 2006; Rowsome 2008). After the revolt, it seems that London recovered slowly until around 80 AD (Rowsome 2008). The work undertaken at Ledenhall Court in the late 1980s clearly showed how this proceeded (Milne 1992, 1995). At the start of the period a number of farms existed in the area that was just outside the city's edge. By 75 AD these were beginning to be replaced by closely spaced urban housing associated with a new road. At this point it seems the town's fortunes changed dramatically and it clearly started to expand at speed (Milne 1995). This phenomenon is also seen at a number of other locations throughout London at this time (Perring *et al.* 1991; Perring and Brigham 2000; Seeley and Drummond-Murrey 2005; Bluer and Brigham 2006). It would seem that that during the period from AD 80 – 100, London became the provincial capital (Mattingly 2006). Of course there has recently been a debate about what we mean by 'provincial capital' with the suggestion made that this is wherever the governor is rather than the paramount economic and administrative centre (Milne 1996; Millett 1998). However, Hassall (2000, 53-54) feels that this is semantics to some extent and pedantic to boot. He also uses the recent translations from the Vindolanda tablets to suggest that that most of the offices of state associated with the governor were permanently located in London. In addition, although the governor may not always have been present, London certainly housed the procurators office (Rowsome 2008). Personally I would prefer my insects to have the status of coming from the capital so of course I fully support Hassall's argument.

Whatever we mean by 'capital', certainly by 80 AD the small 'palace', oft presumed to be the winter quarters of the procurator (Merrifield 1983; Marsden 1980), seems to have been built at the location of Cannon Street station. This is just to the east of the Walbrook and across the stream from the Poultry site. Once again the specific function of this 'palace' has been questioned, with suggestions that it may have been a larger public building, in parts a temple or even a bath complex (Perring 1991; Milne 1995). It has also been suggested that the official residence might have been across the river at Winchester Palace (Perring 1991) where a large stone and tile building has been found dating from 80 – 120 AD (Yule 2005). Certainly the fact that this building is replaced by a far larger palatial building of either military or administrative status early in the 2nd century AD (Yule 2005) is probably of significance. However, Yule (2005) suggests a third alternative. The governor's Palace was probably located near to the Cripplegate fort but has been lost in later Roman re-development of the area, when a palace was built by Allectus in the third century. It is also presumed that the temple to the divinity of the god Emperor was moved from modern day Chelmsford during the early part of the 2nd Century AD. The location of this new imperial cult temple remains something of a mystery, but Marsden (1980) speculates that this could be the small temple found by Frank Cotterill in 1934, next to the forum located in Gracechurch Street; however, this assumption has been questioned by Haynes *et al.* (2000).

What is clear is that during this period we see the construction of a reasonably large basilica (the 'town hall' and civic centre) and a forum (the market) on either side of modern Gracechurch Street where it meets Lombard Street and Fenchurch Street, by 130 AD (Marsden 1987; Hassall 2000; Perring and Brigham 2000; Dunwoodie 2004; Mattingly 2006). This notable structure seems to have been preceded by the smaller 'proto forum' which was established around 75 AD (Marsden 1987; Hassall 2000; Dunwoodie 2004; Mattingly 2006). Other public buildings in the area are two bath complexes at Upper Thames Street and Cheapside dating to later part of the first century (Marsden 1980; J. Hall and Merrifield 1986; Milne 1995). It is presumed that the areas of town between these public buildings contained a mix of shops, warehouses and homes of all statuses (Milne 1995). The waterlogged samples excavated from a series of 'tenement buildings' which developed along the line of a new Roman Road found at Copthall Avenue, in the upper part of the Walbrook Valley, do allow us an insight into living conditions at this time (Malony 1990). Again, they attest to expansion onto previously unoccupied land. Later excavations in the area have clearly shown that the Walbrook valley developed into a pottery production centre (Seeley and Drummond-Murray 2005). There also is a considerable expansion of the area of wharf fronts on the north bank of the Thames at this time (Milne 1993; Perring and Brigham 2000; Rowsome 2008). The district to the west of the Walbrook seems to have been less well-planed with the roads developing outside a regular grid pattern and unevenly spaced, perhaps suggesting that expansion was so rapid that it could not be regulated (Rowsome 2008; Bateman *et al.* 2008).

During the early part of the second century the Roman army was given a new home as well. Where its old home was is still open to debate, but a recent excavation has raised the possibility that a post-Boudiccan fort may have been located near Plantation House (Dunwoodie 2004; Bateman *et al.* 2008; Rowsome 2008). More recently it has been suggested that the billeting of the army may have been scattered throughout the town before 120 AD (Bateman *et al.* 2008). This sizable fort (5 hectares) was at Cripplegate, to the west of the Walbrook and to the north of the Poultry site and was built around 120 AD (Marsden 1987; Perring 1991; Milne 1995; Howe and Lakin 2004; Mattingly 2006; Rowsome 2008). The last

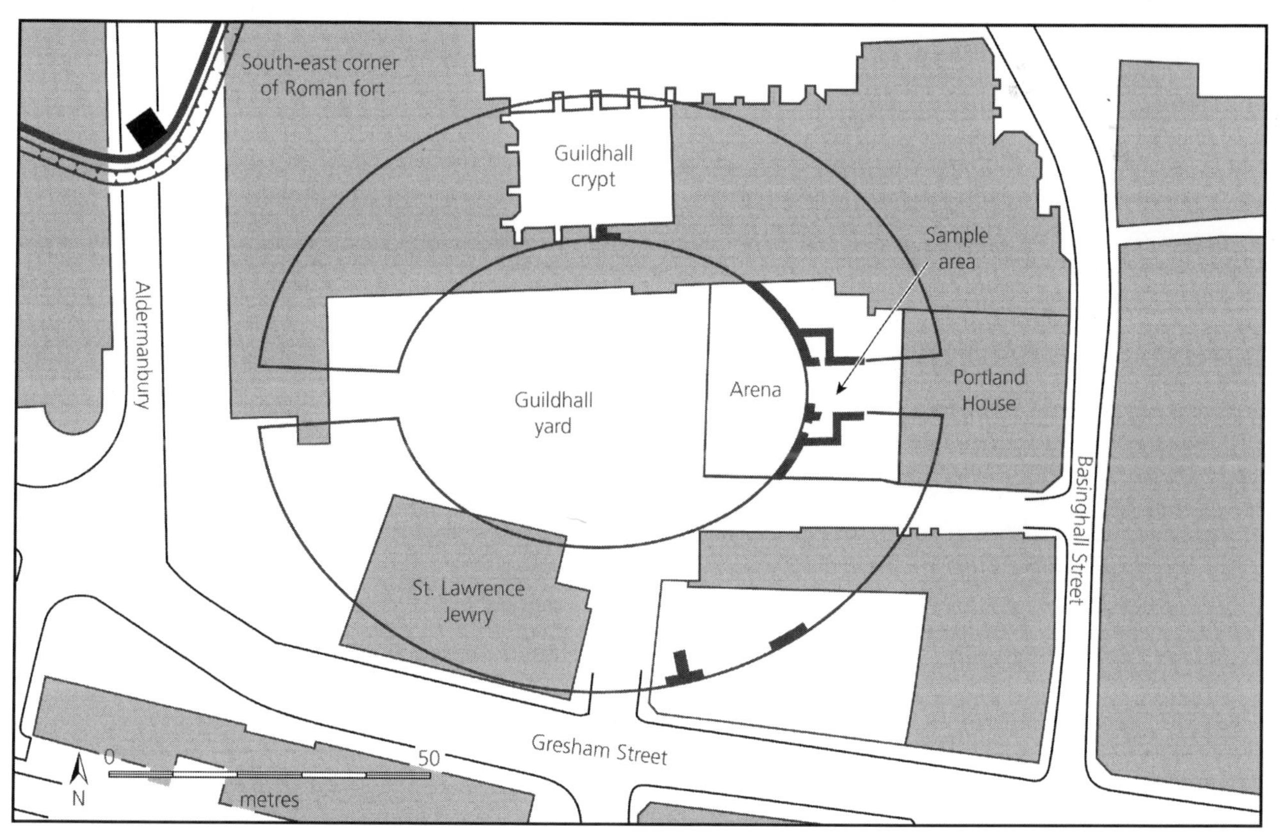

Figure 5.4. The archaeology of the 2nd century Roman amphitheatre at the Guildhall AD 60 (redrawn from Bateman 1997)

civic feature to be discussed here is the Roman amphitheatre found in the courtyard of the Guildhall just to the south of the fort. This impressive part of London's Roman history was only discovered in 1987 (Bateman 1990; Bateman *et al.* 2008). I actually visited this site when I was admitted to the Clothworker's Guild and after I had been given the freedom of the City of London at the Guildhall. At the time I was just starting my Masters degree at Sheffield and the last thing on my mind were insect remains. In fact, at the time, the analysis of beetles was the only part of environmental archaeology that I really hated. If I had realised that eventually I would end up working on samples from the wooden drains I saw at the time, through the builder's shuttering, I might have paid more attention.

There appears to have been two amphitheatres on the site (Bateman 1990, 1997; Bateman *et al.* 2008). The first, dating from within a year or two of 74 AD seems to have been a wooden affair that was replaced by a more substantial structure made partly of stone by the early second century, probably around 125 AD (Bateman *et al.* 2008). The plan of the later amphitheatre is shown in Figure 5.4. This structure was of quite a size and Hassall (2000) and Bateman (*et al.* 2008) suggests that it might have seated between 7000 and 10,500 people. The part of this structure that was revealed during excavation consisted of the eastern 'entrance way' to the arena. A wooden drain that probably emptied out to the Walbrook ran under this entrance (Bateman *et al.* 2008).

THE INSECT FAUNAS FROM LATE FIRST AND SECOND CENTURY LONDON

Only three sets of insect faunas come from sites associated with Roman London have been analysed in the period following the Boudiccan revolt, when the city reached its status as the provincial capital, In terms of chronology the three sites are as follows:

1) a series of housing and road deposits associated with the site at Poultry
2) deposits from occupation horizons at Copthall Avenue
3) drain deposits from the Roman Amphitheatre

If we considered these faunas together, we can gain a general impression of the nature of urban London at this time.

Poultry

After the area's general destruction in the Boudiccan revolt Poultry seems to have re-established relatively quickly (Rowsome 1998, 2000; Hill and Rowsome 2012). Once again a mix of small industry, domestic housing and open yards developed around the crossroads at this time. The layout of the area at this time is shown in Figure 5.5. The ecology of these faunas is shown in Figure 5.6.

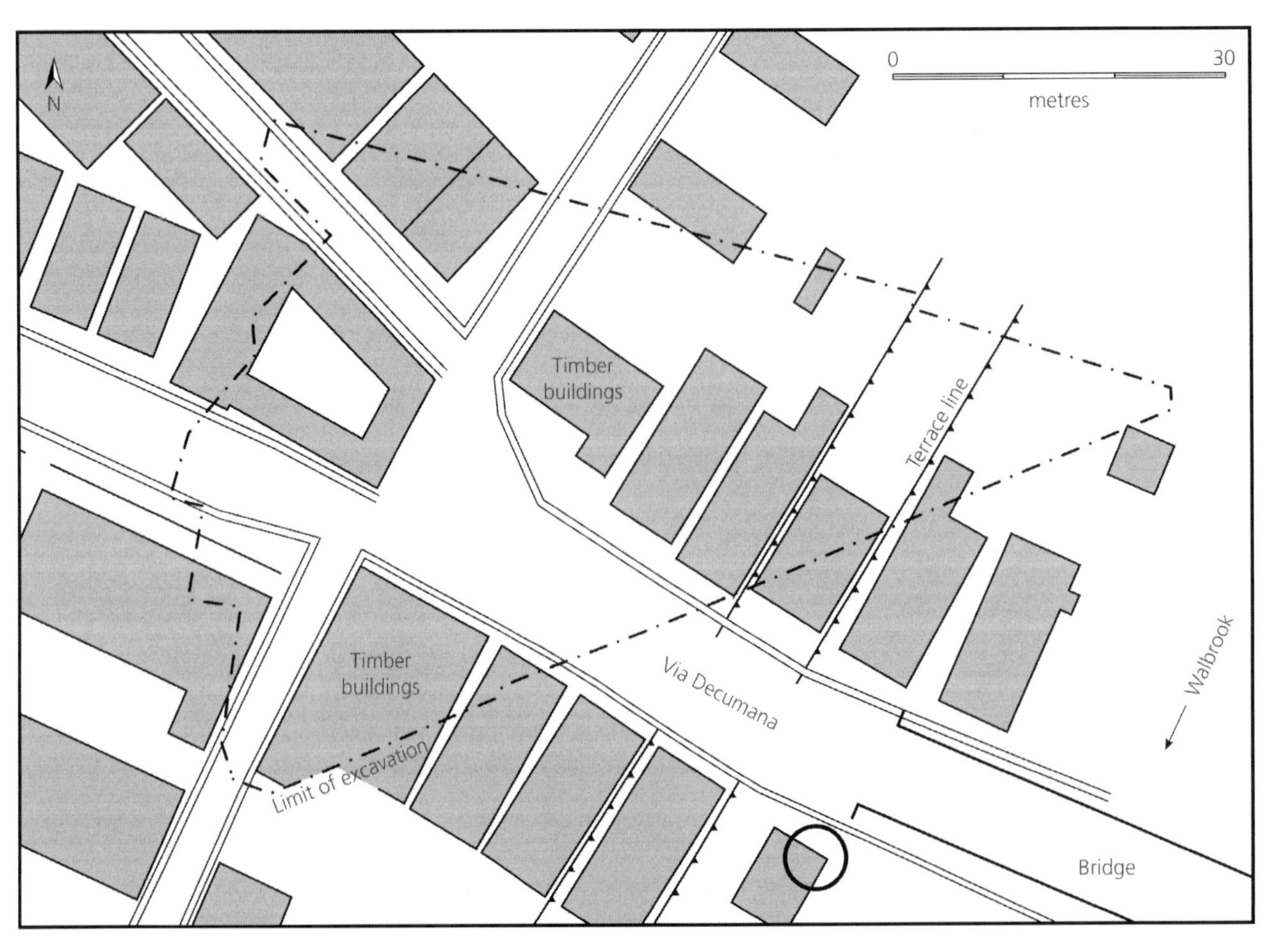

Figure 5.5. The layout of the site at Poultry around 100 AD (redrawn from Rowsome 2000)

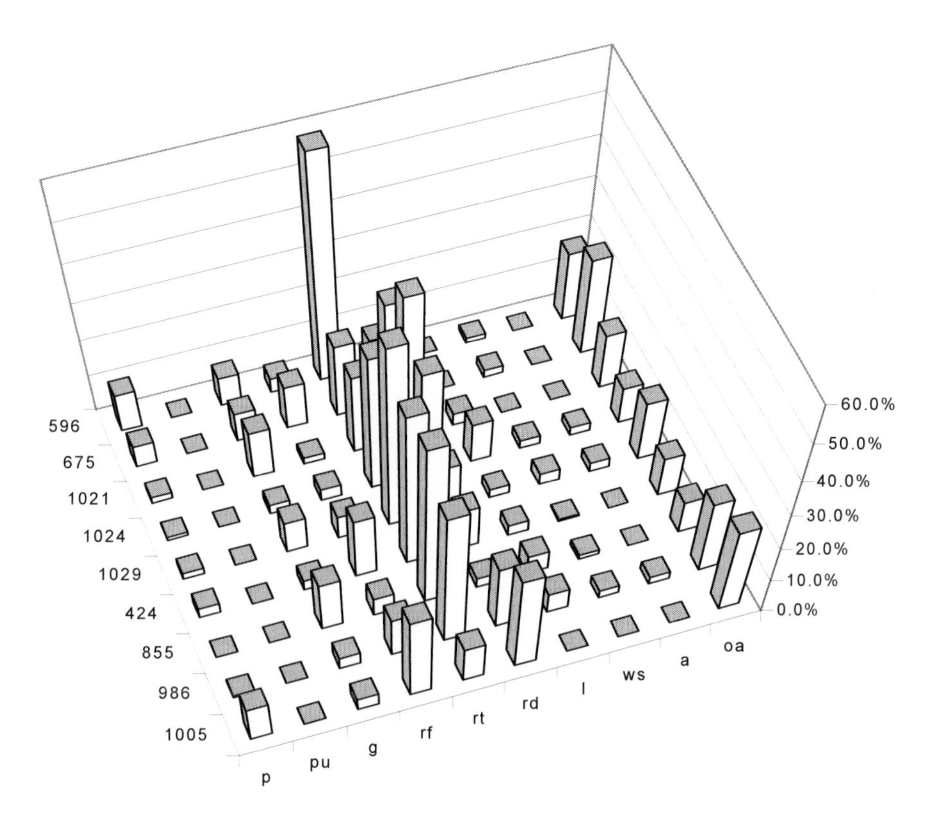

Figure 5.6. The ecological summary statistics for the insects from the late first and second century deposits from Poultry (ecological codes are outlined in Figure 1.2.)

Only two deposits at Poultry containing insect remain (samples 596 and 675 from a well associated with Building 22) come from the period immediately after the Boudiccan revolt (Smith 2012). The fauna recovered gives us an insight into the nature of the yard around the wellhead. Once again the plant feeding species of beetle clearly indicate weedy, open ground though the weevil *Apion difficile* feeds on dyers greenweed (*Genista tinctoria* L.). While it is possible that this was merely growing in the area, it is a common dye plant and may indicate that dying and cloth preparation occurred in the area. However, a large proportion of the species recovered are associated with decaying waste and plant materials from human habitation. This suggests that considerable quantities of settlement waste were either dumped or fell into the well. Indeed one of the examples of the human flea (*Pulex irritans*) from this site came from this deposit. The remainder of the samples from this period at Poultry come from a range of deposits within open areas and yards (samples 1024, 1029, 424, 427, 855, 986 and 1005).

As is becoming a recurrent theme, most of the fauna are species that are associated with mixed rotting and decaying waste and rubbish. Typical of this are taxa such as the various *Cercyon* and histerid species, many of the rove beetles such as the *Oxytelus, Phacophallus parumpunctatus, Leptacinus, Quedius* and *Falagria* species and *Anthicus formicarius*. This waste clearly appears to have an origin in the settlement given the high proportions of 'house fauna' species recovered. Unpleasant conditions are also indicated by several of the species of fly present. These include *Sepsis* species, Copromyzinae, *Ischiolepta cf. pussila* and *Telomerina flavipes*. These flies often are associated with material containing human cess; and Skidmore (1999) holds that the last species may also be indicative of buried material. Samples 424 and 437 seem to have become particularly maggot-ridden, since they contained large numbers of common house fly (*Musca domestica*) and stable fly (*Stomoxys calcitrans*) pupae. The clear impression is that these open yards were covered in scatters of rotting settlement rubbish and cess.

Some of this rubbish appears, again, to have resulted from grain processing, storage or use. Granary weevils and other pests of stored products account for between 5 – 15% of the insect faunas recovered during this period. Though this is not in the same league as the proportions seen before the Boudiccan revolt, grain pests clearly continue to persist. This suggests that large-scale grain storage and processing continued to occur in this area of London.

Drain and open area deposits from the Roman amphitheatre

The insect faunas recovered from the drains running below the eastern entrance to the second century amphitheatre and a number of dump deposits tell a similar story of urban waste (Smith and Morris 2008). The ecology of these faunas is illustrated in Figure 5.7.

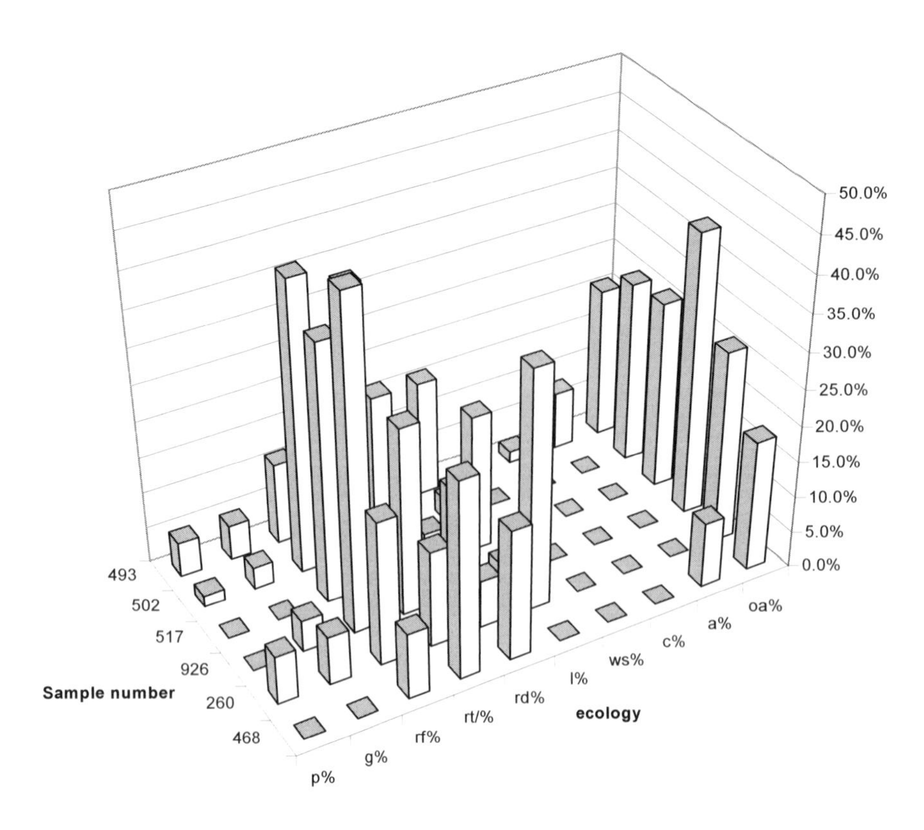

Figure 5.7. The ecological summary statistics for the insects from the late first and second century deposits from The Guildhall (ecological codes are outlined in Figure 1.2.)

The material from sample 493 came from a large wooden silt trap associated with drain S14 from under the entrance way (Bateman *et al.* 2008). This produced the largest and most diverse of the insect faunas recovered from the Roman periods at the Guildhall. The fauna is again dominated by species such as *Cercyon*, staphylinids (rove beetles), cryptophagids and lathridiids; all the classic indicators for deposits of drier settlement waste and rubbish (Smith and Morris 2008). A similar story is told by the plant macrofossils recovered from these drains, which included a range of species associated with rough and wet ground and a small number of food plants such as fig (*Ficus carica* L.) and plum (*Prunus domestica* L.) (Grey and Giorgi 2008). There is a very small collection of granary insects that could have been swept into the drain with other settlement waste, but may also have an origin in cess. The inclusion of grain pests in cess will be discussed below in Chapter 11. The impression is that the drains contained material carried in from nearby surfaces, perhaps from the floor of the amphitheatre itself. The fill of this drain also produced a very large number of larvae of non-biting midges suggesting that the water in the drain may have been relatively still and stagnant (Smith and Morris 2008), these water conditions were independently confirmed by the molluscs recovered (Pipe 2008). Deposits from the other drains (S26, S27: Samples 260, 486), though slightly later in date, show similar materials and conditions.

Three insect faunas were recovered from a range of early second century dump deposits at the Guildhall (502, 517, and 926). These contained the same range of species of beetle as were seen in the drains along with the puparia of the common housefly (*Musca domestica*) (particularly sample 926) (Smith and Morris 2008). One context (502) also produced large numbers of *Sepsis* fly pupae, a species that is often associated with faecal material and cess (K.G.V. Smith 1989). The plant remains from these areas again include seeds of fig, grape (*Vitis vinifera* L.), blackberry (*Rubus fruticosus* L. agg.) and sloe (*Prunus spinosa* L.) (Grey and Giorgi 2008). These seeds are common in cess since they survive passage through the human dietary tract (please feel free to experiment at home – see Osborne 1983). Other plant remains recovered suggest that a rather open scrubby area with weedy plants such as nettle (*Urtica* spp.) and knotweed (*Persicaria* spp.) were present in the area (Grey and Giorgi 2008).

All in all, it looks as if areas of the arena and the entrances to the amphitheatre contained discarded, or washed in, waste and rubbish. This is not, of course surprising. There is evidence for the arena floors being made in part of *opus signinum* (a flooring consisting of rough gravel, stone and ceramic waste cemented into place with mortar) (Bateman *et al.* 2008); however, this surface appears to have had covered with layers of silt and sand (Bateman *et al.* 2008). It is quite likely that settlement rubbish and waste would have become incorporated quickly into these silty/ sandy layers above the hardened *opus signinum* amphitheatre floor. It has also been suggested that organic matter and ashes may have been incorporated deliberately to 'soften' the arena floor (Bateman *et al.* 2008). Insect remains suggest that almost any open area or surface in Roman London seems to have accumulated dumps of urban waste at some point.

15–35 Copthall Avenue

This site lies to the north of Poultry at the upper reaches of the Walbrook Valley. The insect remains were studied by Enid Allison and Harry Kenward, the plant macrofossils by Anne Davis and Dominique de Moulins and the pollen by Rob Scaife. The results of this work are presented and discussed by de Moulins (1990). During the later first millennium, the site had been covered by a number of drainage channels that are thought to have been associated with a programme of reclamation and land levelling in this area (Maloney 1990). All environmental indicators from these ditches, particularly the pollen, clearly suggest that an area of swampy ground and damp grassland predominated. Over time the area seems to have become more 'disturbed' with both pollen and plant remains suggesting a flourishing growth of chickweeds (*Stellaria* spp.), knotgrasses (*Polygonum* spp.) and nettles (*Urtica dioica* L.). The insect and plant remains also suggest that settlement waste and cess were entering these channels, even early in the period. Small quantities of insect grain pests were recovered as well. One sample from this period, associated with the road that crossed the area, also contained large numbers of the eggs of the intestinal parasitic worm *Trichuris trichiura*, suggesting that cess had been deposited in the area. The impression gained is of a damp and swampy area where rubbish was either dumped or washed in.

By the 2nd century buildings were beginning to encroach into this area, again indicating the rapid expansion of the town at this time (Malony 1990). The palaeoenvironmental evidence from a range of floors, pits and gullies associated with buildings suggest that though the area remained low-lying and swampy, considerable amounts of the usual range of Roman waste and rubbish accumulated.

LATER ROMAN INSECT FAUNAS

Only two deposits from the 3rd and 4th centuries of London have produced insect faunas. These are from the later phases at Poultry and Copthall Avenue. At both sites the deposits are associated with roads and the fills of roadside gullies. The ecology of the faunas from this period at Poultry is illustrated in Figure 5.8.

Sample 718 at Poultry, from a dump deposit sealed by the construction of the road, is dominated by species associated with grain and grain storage (70% of all taxa) (Smith 2012). The vast majority of these are 'the granary weevil', *Sitophilus granarius* (104 individuals). This probably indicates that whole grain, in the early stages of its decay, was incorporated into the dump of destruction debris associated with road construction.

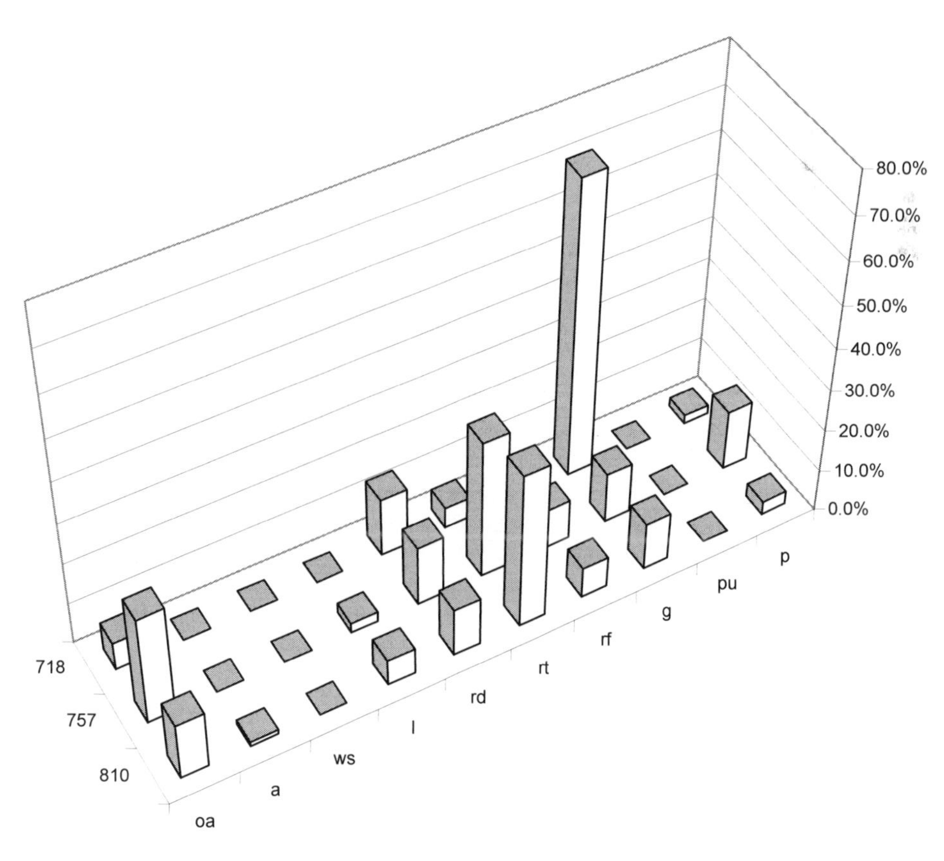

Figure 5.8. The ecological summary statistics for the insects from 3rd and 4th century deposits from Poultry (ecological codes are outlined in Figure 1.2.)

Samples 757 and 810 are again associated with road construction. Though only small insect faunas were recovered they indicate that relatively foul domestic and settlement rubbish was incorporated into these deposits. Sample 810 seems to have become particularly maggot-ridden given that thirty-two individuals of the pupae of the common house fly *Musca domestica* were present.

At Copthall Avenue the insects (and plants) from a series of roadside ditches suggest that this area remained very damp and muddy. There is evidence, mainly from the insect remains and the internal parasite ova, that settlement waste and cess continued to be dumped in the area or washed in from the surrounding town (de Moulins 1990).

COMPARISON WITH OTHER ROMAN SITES

The insect faunas from Roman London have given us a fairly clear picture of the nature of urban life in the city. In general most deposits appear to be collections of settlement waste, food debris and cess. The impression gained is of a busy city where the roads, buildings and yards were often covered in scatters of decaying plant material, waste from industry and the general filth of living. Cess and rubbish disposal also seem to have been rather haphazard. In the first chapter I did alert the reader to my worries that these insect faunas, and the apparent story they tell, could result from the reburial of material, post-depositional decay and so on. However, the impression of general squalor is so ubiquitous that there must be some truth to it.

Similar faunas and environmental conditions are also seen at a range of other archaeological sites from the Roman period. Of course the main comparison to be drawn is with Kenward's work from the 2nd and 3rd century Roman Colonia at York (Hall and Kenward 1990). This work consisted of over 500 samples excavated from a similar range of buildings to those seen at Poultry. Despite the larger dataset from York there is essentially a clear similarity between the two sets of insect faunas. In terms of the range and relative proportions of the synanthropic, 'house fauna' and decomposer communities present the two datasets are almost interchangeable. The main difference is the dominance of grain pests in some of the samples from Poultry, and the occurrence of an insect fauna at York that clearly suggests the inclusion of stable manure in some deposits. Though there are suggestions that stable waste may have been present at Poultry, particularly the odd find of *Damalina* lice, it does not appear to have been an important component. Minor differences such as these between the sites probably relate to the local issues of building use rather than any intrinsic differences in terms of formation or function between York and London. Indeed, Kenward has suggested that:

"Roman 'urban' settlement produced rather similar conditions wherever it occurred, in England at least"' (Hall and Kenward 1990, 393)

None of the sites from London discussed here contradict this view. In addition to York, the same 'urban fauna' has also been recovered from both the town and the fort at 1st and 2nd century Carlisle (Allison *et al.* 1989; Kenward *et al.* 2000; Smith 2010), 2nd century Alcester (Osborne 1971, 1994) and Chichester (Girling 1989b).

There are, however, some specialised 'dump deposits' from Roman London that stand out from the general picture of usual urban muck. In particular, these deposits seem to be associated with isolated attempts to dispose of rotten grain. Examples from Roman London are the deposits associated with 1st century Building 4 at and the road 'packing' or 'levelling' deposits from the late periods at Poultry. Why this activity occurred and why it is important will be discussed in detail later in this chapter, since it does seem to commonly occur at several Roman sites in London.

THE DEVELOPMENT OF 'URBAN FAUNAS'

Questions that should be asked at this point are: Where did this urban fauna come from?'and Where has it gone?

Kenward and Allison (1994) tackle these points in an interesting manner. Of course the first issue raised is that we should not regard this 'synanthropic' or 'urban fauna' as ...urban. They establish from the start that the species recovered from urban archaeological sites all occur in nature in the British Isles. They, therefore, suggest that the original habitat for these species is out in the countryside. For example:

1) Species that are commonly associated with dried hay and other plant matter in the archaeological record (e.g. *Typhaea stercorea, Cryptophagus spp.* and *Lathridius* spp.) are also found in dry material, dead grass and plant litter in a wide variety of situations ranging from the inside of hollow trees to the base of hedge rows.

2) The small rove beetles, *Trogophloeus, Platystethus* and *Oxytelus* species that commonly occur in a variety of soft and wet archaeological yard and floor deposits today are also frequently found around muddy and plant strewn watersides.

3) Many of the members of the 'decomposer community' associated with rotten waste and urban filth from archaeological sites are, today, either associated with animal dung or wet leaf litter found in open country.

4) Several species that archaeologically are seen as extremely synanthropic such as the spider beetles (*Ptinus fur* and *Tipnus unicolor*), carpet beetles (species of *Anthrenus* spp. and *Attagenus pellio*), mealworms (*Tenebrio* spp.) and some hide beetles (*Dermestes* spp. and *Trox scaber*) are also found today in birds nests, hollow tree stumps or dried corpses.

There also are very few species amongst this 'urban fauna' that do not appear to be present as 'natives' in Britain. Certainly results from Early Holocene sites include many of these species, albeit in low numbers.

Kenward and Allison (1994) suggests that the 'urban fauna' probably came together as a set of 'communities' very early in prehistory, perhaps as humans moved into permanent settlements and started to farm. It seems that our activities often offered these species just the type of habitats and conditions they needed. Given that large populations of these species seem to occur in archaeological contexts it is possible that we actually presented them with ideal conditions with much more abundance and regularity than they could possibly find in the wild. Kenward therefore sees the first step in the trail towards synanthropy (i.e. being pre-adapted to forming a 'relationship' with human settlement) resulting from species either slowly making their own way into small farming settlements or arriving in material brought in for use in building, fodder, fuel or craft materials. Certainly, small farm sites and settlements in the archaeological record contain most of the 'urban fauna' recovered from London (Kenward and Allison 1995; Girling 1979c; Smith *et al.* 2000). It also is clear that the formation of these insect 'communities' did not wait for the Romans; they have been recovered from Bronze Age settlement materials at the Mere Lake villages (Girling 1979c) and Iron Age Goldcliff, Gwent (Smith *et al.* 2000). Various members of this community also have been recovered in Neolithic deposits in northern Europe (Edith Schmitt *pers. com.*), Bronze Age Greece (Panagiotakopulu 2000) and from Pharaonic Egypt (Panagiotakopulu 2000, 2001). However, the number of small farm sites studied to date is actually extremely limited. An interesting question for the future will be to try to establish if there is a distinct difference between the insect faunas of town and farm and, if so, to establish why this might be.

So, in summary, we should not regard these species as really 'urban'. They are 'country cousins' who continued to behave in the same way in town as they did in the country; they just did so very successfully. This is of course a process that can be repeated even on the larger scale represented by a city. (If we wished to be facetious, as with that other pest of urban life, the cat, we could ask who gains most from this relationship?)

So where has this 'urban insect' fauna gone? Kenward and Allison (1995) clearly lay the blame on changes in the way we live over the last 50 years. Central heating, the vacuum cleaner and changing perceptions about personal and housing hygiene are all considered to be responsible for the decline of the 'house fauna'. Major factors could certainly include decreasing use of thatch roofs, earthen floors and cesspits. A sad picture, therefore, can be painted for the six-legged former masters of the urban environment being slowly pushed towards extinction. If they avoid this fate, at best they will be relegated to the garden compost heap. Shed a tear for the human flea. It is described as 'disturbingly rare' in Britain (interesting to note that there are no campaigns in the press to save this seriously endangered species). We face a modern world of glass and concrete where our only insect friends are woodworms and carpet beetles. I am not quite so pessimistic. One of the advantages of teaching 18 year old University students is you periodically get to see the standard of housing, and the way, in which they live. As discussed earlier in the infamous grill pan incident, much of student housing serves as a refuge for many members of the 'house fauna'. Students, be proud! You are doing your bit for nature conservation.

GRAIN PESTS, STORAGE AND THE ROMAN ARMY

Amongst the most striking finds from the pre-Boudiccan areas of the Poultry site were the two faunas that were almost exclusively composed of a range of granary pests. In addition, the grain pests from other samples at this site often account for more than 20% of the total fauna.

At a primary level this clearly does indicate that storage and use of grain occurred on site. It also probably confirms the suggestion made by the archaeologists that the 'timber post structures' on site were granaries. However, there is a lot more to discuss here. This discussion takes the interpretation of these faunas beyond mere 'landscape reconstruction' or 'deposit identification' towards the interpretation of human behaviour and patterns of life. For example the following questions could be asked:

1) Is this abundance of grain pests seen elsewhere in Roman Britain or is Poultry an exception?

This question has recently been addressed in detail in a paper by Smith and Kenward (2011). This dominance of grain pests is not an unusual phenomenon for the Roman period. This is illustrated in Figure 5.9 where the relative proportion of grain pests in the insect faunas from a number of 1st and 2nd century sites is illustrated. Amongst these are the warehouses at Coney Street, York. This is the most spectacular example of the archaeological occurrence of grain pests. Four small samples were collected from silts taken from the beam slots beneath the floor of these warehouses. Over 93% of the insects recovered were grain pests (Kenward and Williams 1979), suggesting this warehouse was very heavily infested. Another good example is that of the 2nd century well from just outside the small legionary fort at Invereskgate, Scotland. Here the grain pests in eight samples from the bottom of the well accounted for 60% – 70% of the insect faunas recovered (Smith 2004). Deposits associated with the yards and possible stable blocks in the various first century phases of the military fort at the Millennium site in Carlisle also contain relatively large proportions of grain pests (Smith 2010). A similar dominance by grain pests is also seen in the 2nd and 3rd century wells at Skeldergate and Bedern, York (Hall *et al.* 1980; Kenward *et al.* 1986) and similar proportions are also commonly found in the numerous deposits from the Colonia at York (Hall and Kenward 1990).

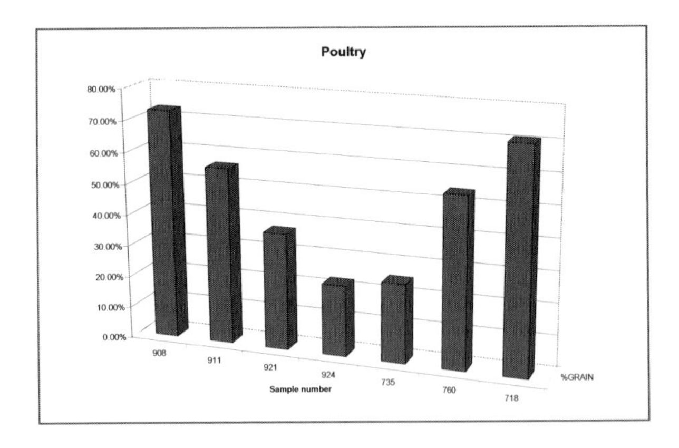

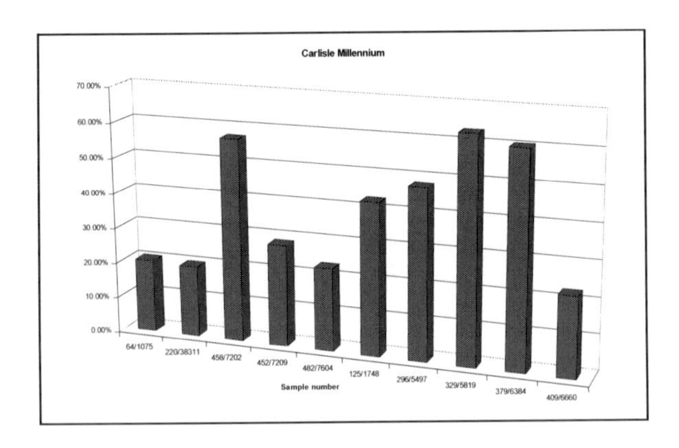

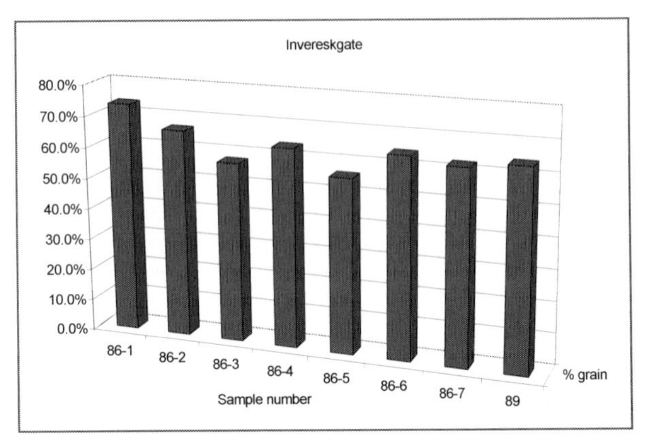

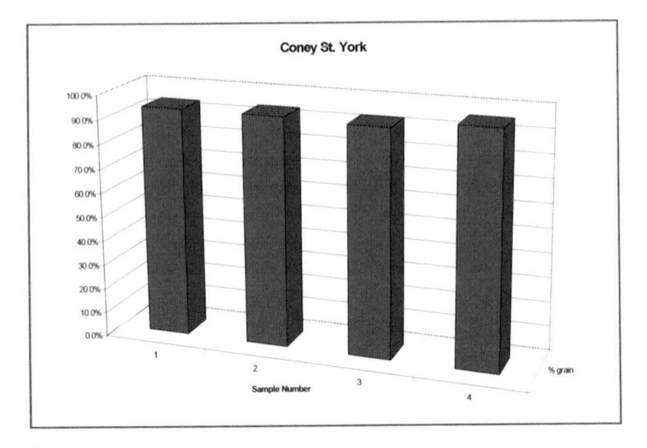

Figure 5.9. Selected samples that are dominated by grain pests from a number of Roman sites

This data leave us with the impression that grain pests were relatively common, if not dominant, part of the insect fauna of Roman settlement. Now, before everyone assumes that grain pests swamp all Roman deposits, a quick review of the literature does suggest an additional pattern. Many rural sites **do not** appear to produce large faunas of insect pests. Perhaps, this is a problem associated with towns and larger settlements. However, there is an additional issue. Very few of these rural insect faunas are from farm buildings; rather they are from field ditches and wells. At present, therefore, we run the risk of comparing apples with oranges. It is also noticeable that many of the sites where grain pests are dominant in the insect faunas also appear to have connections with the Roman Army.

One clear exception to this urban rural divide is the recently excavated first century site at Northfleet, Kent. Here both the charred deposits of malting waste and the fill of a large timber-lined cistern contained insect grain pests (Smith 2011a). It is felt that malting of grain for beer was common on site and the malt itself was probably infested (Smith 2011a; Smith and Kenward 2011, 2012). Whether insect infestation may have tainted the taste of the beer is not clear, but the idea of using 'living grain' is a bit stomach churning.

2) How do these grain pests get into the archaeological record in such numbers?

This is a subject that has been discussed in a recent paper by Smith and Kenward (2012). We can easily explain away small numbers of grain pests in settlement faunas: it can result from small amounts of spoilt grain or foodstuffs being included in the disposal of household rubbish. It also is possible that some of these species could make their own way out of grain stores into nearby deposits.

Another common suggestion for how grain pests found there way into urban deposits is that they became incorporated into food, perhaps in pottage, gruel or rather unrefined bread, and subsequently pass into cesspits where small numbers of these species are common. Peter Osborne (1983) set out to prove that the human digestive tract could act as a route for the incorporation of insects into such deposits. In particular, he wished to prove that they could survive the journey. This resulted in one of the most spectacular biological experiments of our time. He took 20 granary weevils and 20 saw toothed grain beetles, put them in his morning porridge and collected what came out the other end 24 hours later. (The usual practice in such experiments is to eat some sweet corn at the same time since it passes through your intestines largely undigested and, therefore, ensures you find the right bit). Back in the lab, presumably to the surprise (? disgust) of his co-workers, he processed 'the material' in the usual way. Thirty-seven undamaged individuals were found, with only three getting lost along the way. The thirty-seven recovered beetles had disarticulated; presumably the connective tissues were digested by Peter himself, but appeared to have suffered no physical or other damage (Osborne 1983).

I also have found another rather surprising route by which the grain species can be added to archaeological deposits in small numbers. A few years ago John Letts of Reading University asked me to examine the insects from 14th century AD smoke blackened thatch (Letts 1999). More will be said about this incredible substance in Chapter 10, but in this case the cereal thatch contained a lot of loose grain and this clearly attracted granary weevils (Smith *et al.* 1999). It is therefore possible that where cereal thatch is incorporated into archaeological deposits we should not be surprised if granary weevils also are present.

However, this kind of casual incorporation cannot explain deposits where the species are dominant and found in very large numbers. This needs a much more deliberate explanation. By deliberate I do not mean my approach to interpretation but rather the way in which humans behaved in the past. Here we have ancient behaviour and decisions being made that directly effect what we find in the archaeological record. In this case it mostly centres on attempts to get rid of rotten and infested grain.

There are actually only a limited number of things you can do to get rid of *Sitophilus granarius* and its chums once they are in your grain. The key thing is to make sure that they cannot infest more grain by making sure you kill the little blighters with supreme efficiency and ruthlessness.

Fire seems to have been one rather drastic technique used. There is of course the famous case of the Malton Burnt Grain. When Corder excavated the ditches of the fort at Malton, North Yorkshire, in the 1920s he found a 9 metre wide and 300 mm thick deposit of burnt grain. Over the years this was explained away as being the result of the storming the fort by the Gallic invaders after the withdrawal of the legions by Rome around 295 AD. When Paul Buckland (1982) examined a small sample of this grain he found numbers of burnt granary weevils, the saw toothed grain beetle and the rust red flour beetle (*Cryptolestes*). Obviously the solution chosen by the army in this instance was to take the stuff away from the grain store and set it on fire.

Throwing rotten or infested grain down wells also seems a popular disposal technique. This can be seen clearly at Invereskgate (Smith 2004), Skeldergate, York (Hall *et al.* 1980) and Bedern, York (Kenward *et al.* 1986). At Invereskgate large numbers of midge larvae suggested that the water had 'soured' before disposal (it does seem sensible to use a well that has gone out of use for disposal rather than one you are still using). In fact, the Roman's/ Romano-British seem to have been quite keen on throwing rubbish down disused wells. When I examined the well deposit form the Mancetter Roman fort in the West Midlands a few years ago I came to the conclusion that, in this case, stable waste had probably gone down (Smith 1997a). In fairness, it must also be said that the Romano-British also seem to have been keen on throwing various votive offerings, dead animals and bits of humans as 'closing deposits' into wells as well (Esmonde Cleary 2000; Woodward and Woodward 2004).

Another solution would be to bury the deposit deep enough to stop the insects getting out again. This seems to have been attempted at the Coney Street granary. Here a layer of silt seems to have been used to seal the infested area when the granary was rebuilt (Kenward and Williams 1979). This is also probably what happened in the later periods at Poultry with infested grain being 'tucked away' into the levelling deposits for the new roads.

Another solution would be to feed the rotten grain, and therefore the grain pests, to livestock. Kenward and Hall (1995; 761) have suggested that infested grain was probably commonly fed to horses during the Roman period. This is probably the explanation for the large numbers of grain pests recovered from the drains below the possible cavalry barracks at Carlisle Millennium (Smith 2003), as well as elsewhere in the town (Kenward *et al.* 1992a, 1992b, 1992c).

Finally, quite what we are to make, except for a vague gagging sound, of the use of infested grain to make beer at Northfleet Roman Villa is not clear. However, the possibility that a certain level of grain infestation was simply tolerated in the past seems plausible. . In the 1940s – 1960s low levels of grain infestation were tolerated by malters/ brewers, who would only deal with infestations if the grain was 'swarming' with insects or infested by mites, which will taint the taste (Hunter *et al.* 1973). As a small child, my wife told her grandfather that there was a grub in her ear of sweet corn that she was eating. Her grandfather, who had experienced the US Great Depression, replied, 'don't worry, it's just a bit of protein!'

3) Would the presence of large faunas of grain pests have implications for the Roman Army and agricultural production in the period?

I must confess to liking the image of the efficient and ruthless killing machine of the Roman Army spending a lot of their time running around finding places to get rid of infested grain. I am reminded of some of the Asterix books where the legionaries looked and behaved like Laurel and Hardy. Certainly such major infestations must have represented a considerable nuisance for the officers and a major obstacle to supply and provision for the average legionary. If we take the evidence at face value this may also have been quite a common occurrence. At the level of the wider population the depredations of these creatures may have had a wider impact. In 1978, Paul Buckland made a number of rather pithy and important points concerning implications of granary pests for reconstructing Roman agricultural yields and possible population. He cited a the United Nations Food and

Agriculture Organisation report that concluded that at least 10% of the world's cereal production was lost to insect attack at that present time. Potentially this could have been a much higher proportion in the past. As a result, he suggested that any relatively simple or direct correlation between the amount of grain produced and the 'number of mouths fed' by the Roman Empire was essentially flawed (Buckland 1978, 43). Buckland's *coup de grace* of a conclusion was that:

> *If this hypothesis is correct, the great increase in land under cultivation [in Britain] sometimes claimed for the Roman period may not be the result of increased population under the* Pax Romana *or the heavy burden of the* annona militaris *but the outcome of the increasing attention of an unwanted guest,* Sitophilus granarius *[the granary weevil], whose activities could have accounted for well in excess of 10% of the cereals produced in the Lowland Zone'* (Buckland 1978, 45).

However, one must ask whether this problem only occurred on such a scale in Britain? It is tempting to suggest that this may have been simply a local factor, resulting from the attempt to keep grain dry, and therefore stop insect infestation, by using techniques developed in the Mediterranean and that these were not adapted to the rigors of the British climate. However, there is no evidence to support this conclusion. Archaeoentomology is essentially a North European phenomenon and we have no comparable insect faunas from the rest of the Roman Empire. However, Panagiotakopulu (2000, 2001) has clearly indicated that these species presented similar problems in Pharaonic and Byzantine Egypt. It seems likely, therefore, that granary pests would have affected the rest of the Roman Empire as well.

4) What is the origin of these species in Britain?

Now we hit an interesting and curious fact. There has never been a single find of a grain beetle in Britain before the Roman invasion (Buckland 1978; Kenward and Smith 2011). This is despite a considerable number of pre-Roman sites having produced insect faunas. Buckland addressed this problem in his 1978 paper. A key factor, he suggests, that stops the spread of these pests to Britain, is that before the Roman invasion most grain was stored below ground in pits. It is known that from the work of Dendy and Elkinton (1920) on sealed storage, and the results of the experimental storage pits at Butzer (Reynolds 1974) that high levels of carbon dioxide build up in such pits. A second factor, often suggested by students, is that grain production and consumption may well have been local with little trade in grain between settlements and across the channel. Later Buckland verbally referred to pit storage in particular as the '*cordon sanitaire*' that stopped the spread of the grain pests. There are of course objections to this argument. What about the prevalence of 'four post' granaries throughout Iron Age Britain? Are these actually grain stores? It is also suspected that grain was traded widely, and across the channel throughout the Iron Age (Mattingly 2006). However, despite such sensible objections the fact still remains that there are no grain pests before the Romans. (I hereby make a rash promise to the next generation of archaeoentomologists; find *Sitophilus granarius* in large numbers in pre-Roman deposits in Britain and I will buy you a bottle of single malt whiskey).

So we could see the British dealing not just with the detrimental impact of Roman invasion after 43 AD but also increased loss of arable production resulting from the first occurrence of grain pests in Britain. Certainly the results from Poultry (Smith 2012), Carlisle Millennium (Smith 2010) and Ribchester (Buxton and Howard-Davis 2000) show that grain pests are there right at the start of the Roman occupation. I suppose one answer to the proverbial question *'what did the Roman's ever do for us?'* Besides the usual retorts of *'roads, baths and* aqueducts'; we can now add *'grain pests!'*

CHAPTER 6: THE USE AND ABUSE OF 'INDICATOR SPECIES'

One of the weapons held in the armoury of any archaeoentomologist is the ability to make a strong deduction about the nature of a deposit based on the presence of a single or small group of species within a much larger fauna. The logic for doing this is quite straightforward. Sometimes a particular species, or group of taxa, are so strongly associated with particular conditions or a habitat that it can resolve the interpretation of a deposit or an archaeological site completely. Usually the importance of these 'indicator species' is also determined by their ability to address a particular archaeological argument. Often there is one particularly significant species that can produce a result that is revealing in terms of human behaviour. How this is done is probably best seen by discussing some 'classic' examples.

Granaries and grain pests

Grain pests are the primary example of this kind of very straightforward interpretation. Much of the discussion of grain pests in Chapter 5 relied on their very specific ecology and allowed us to use them as 'classic indicators'. That one small beetle can indicate so much about human life shows something of the strength of this approach.

Indicators for woodland

In Chapter 3 I discussed two species of insects that seem to be particularly revealing concerning the nature of early Holocene woodland. There was *Ernoporus caucasicus* that suggested lime trees had a stronger occurrence in the ancient woodlands of England than indicated by pollen, and *Scolytus scolytus* with its involvement in the discussion of the 'elm decline'. These are of course classic examples of the use of 'indicator species' to shape the interpretation of very large insect faunas.

Dung beetles as indicators for pasture and grazing

Often finds of large numbers of dung beetles can be taken to indicate the presence of pasture and grazing (see brief discussion on this in chapter 3). A classic example of this is the interpretation of the 'shaft' at Wilsford, Wiltshire. This was a 30 m deep shaft cut down through the chalk of the downs which was originally early Neolithic in date and then recut in the Bronze Age. At the time of excavation, and subsequently, this was thought to be a 'ritual' shaft (see Ashbee *et al.* (1989) and Darvill (2006) for an outline of this argument). However, Osborne (1969, 1989) found that the insect fauna from the base of the shaft was dominated by large numbers of 'dor' beetles (*Geotrupes* species) and 'dung' beetles (*Onthophagus* and *Aphodius* species). This lead Osborne to suggest that the shaft was in fact a well used to water the large herd of cattle that was obviously present around the top of the shaft. This, and the presence of the remains of a wooden container in the shaft, has lead to a drift in the interpretation of this feature from the prosaic to the pragmatic and, perhaps, rather mundane. This rather boring view of the feature is a touch unfair. It is not clear if the insect fauna relates to the original shaft, cut in the forth millennium BC, or to the later second millennium recut which may have occurred for a completely different reason to that of the earlier shaft (Darvill 2006). Even if the shaft is 'merely' a well, its construction involves considerable labour and investment and must have some significance and meaning for the community at the time.

Another classic example of the use of dung beetles to interpret archaeological deposits is the insect faunas from the salt marsh palaeochannels that surrounded the Iron Age buildings at Goldcliff, Gwent (Smith *et al.* 2000). Here they clearly indicated that a large herd of cattle was present at the time the buildings were in use. The presence of cattle was also confirmed by finds of the cattle louse (*Damalina bovis*) and mites associated with animal dung in the buildings themselves (Schelvis 2000), and the presence of considerable numbers of cattle foot prints. Bell (Bell *et al.* 2000) has interpreted this as evidence that the people of this area of the Severn Estuary regularly had livestock graze the foreshore in the summer.

Insects and plants linked to particular craft activities

Indicator packages, including insect remains for craft/ industry have been widely discussed by Hall and Kenward (2003). They outline a range of species, both insect and plants, associated with tanning, textile working, butchery and meat processing and other crafts. The classic example of this is work associated with dyeing at 16-22 Coppergate at York where the plant remains of dryer's greenweed (*Genista tinctoria* L.), madder (*Rubia tinctorum* L.) and woad (*Istatis tinctoria* L.) were all found along with weevils associated with *G. tinctoria*. Given the history of these plants, and that they occurred in large quantities in the deposits behind the tenements, the interpretation was straightforward. The tenements in this area were, amongst other things - dying works (Kenward and Hall 1995). It should be noted that Hall and Kenward (2003) intend these indicator groups to be used as part of larger 'indicator packages' that pull together botanical, entomological and archaeological criteria (see Chapter 10). However, the temptation is always there to push one line of evidence on its own. This is particularly true when working with very disturbed deposits or at the site level (see Chapter 8).

Ectoparasites

These can be used to indicate which stock animals were present on site. This normally consists of looking for various species of lice. These splendid animals are usually associated with a specific host. Sensibly enough, this is reflected in their names. For example, *Damalina ovis* is the sheep louse, *D. bovis* the cattle louse and *D. caprae* the goat louse. Just to break this cosy pattern the pig louse commonly found in the archaeological record is called *Haematopinus apri* and is normally associated with wild boar. This is not the same species of louse that is associated with the domestic pig today which is *Haematopinus suis.* This has led Kenward and Hall

(1995, 778) to suggest that the pig of the past may have strongly resembled – or originated from - the wild boar. One problem that has been raised with using these species to indicate the presence of domesticated animals is that it is possible they came in on hides rather than the live animal (Kenward and Hall 1995, 778; Hall and Kenward 2003).

There are some classic examples of using parasite remains as 'indicator species' to suggest human behaviour. For example when Paul Buckland examined a 3.5 kg sample (relatively small by archaeoentomological standards) within the drain from below a floor at the 17th century AD farm of Stóraborg in Iceland (Buckland and Perry 1989), he found 181 individuals of the 'sheep ked' *Melophagus ovinus* and little else. This biting fly burrows into the fleece and skin of sheep where it feeds on their blood. It is this insects that give rise to 'scab' attacks in sheep. It must also be said that it is a considerable, and painful, nuisance to humans undertaking field survey in Northern Britain (and of course to the dogs that go with them). The deposit also contained 22 individuals of the 'sheep louse' *Damalina ovis*. Buckland and Perry's (1989) interpretation of how such high numbers of these species occurred in the drain is a splendid example of intuitive logic. Paul Buckland suggested, in a paper elegantly named '*Piss, parasites and people: A palaeoecological perspective*' that this results from wool processing. A common stage in work with sheep fleeces is to remove the grease from the wool before it is carded and spun by submerging it into an ammonic and strongly alkaline liquid. In this case he suggests it is soured urine. He relates that in recent history urine was collected throughout the summer in half barrels set into the floors of Icelandic farms. From this he goes on to suggest that the white deposit in the barrels found during the excavation of the Ströng farm (Rousell 1943) was not '*skyr*', fermented yoghurt, but rather the remains of stale urine. From this Buckland goes on to question the common assertion in Norse archaeology that finds of half barrels buried into the floor of a room indicate that it was a larder, by suggesting it was used for wool processing and cleaning.

Another good example of using parasites to suggest the use of a building is that of Schelvis (2000) and his finds of cattle lice from the floor deposits within the Iron Age buildings at Goldcliff, Gwent.

THE ABUSE OF INDICATOR SPECIES

Now, all of these examples show the strength of using indicator species. They clearly can help you to interpret deposits with some degree of confidence. They often are a kind of intellectual 'golden egg'; finding species like these on your site is the lucky deflection that lands the archaeological ball in the back of the goal. Quite often their presence leads directly to a very interesting story of human life. However, the main failing of 'indicator species' is that they can lead to false starts and mistakes in interpretation. For example, in Chapter 3 I discussed the role of dung beetles in terms of indicating clearings and large herds or grazing animals in woodland. It was suggested that, for a number of ecological and depositional reasons, you could not really use these species to answer that question in such a direct way. The problem was the attempt to use dung beetles as an 'indication' for a specific interpretation without really thinking through the nature of the connection between the two. In this case the relation between numbers of dung beetles and area of clearing actually was quite shaky.

The other problem is a very human one. It is a crime against archaeoentomology specifically, but archaeology in general, called 'overemphasis'. This is a common mistake seen in many an undergraduate (and post-graduate) dissertation. The logical extreme of this approach is that every species in a fauna can have a unique significance that fundamentally changes the nature of the site. This leads to contradictory interpretations of the same deposit. This mistake is often compounded by putting emphasis on the 'wrong' bit of ecological interpretation. For example I had a student a few years ago who used the excellent BUGS programme (Buckland and Buckland 2006) to find out about the ecology of the species she had from an archaeological trackway. The student duly found out that *Enicmus minutus* (a plaster beetle) often occurs in stored hay. The student, therefore, suggested that large quantities of hay were carried along the trackway on the strength of one beetle. The problem of course is that this species occurs widely in nature, and is commonly recovered in nearly all archaeological deposits.

Another classic example of this is from a colleague of mine who worked on trackway deposit for a certain television show. Over the course of three days the director and presenter kept asking if he had found any dung beetles. Of course what they wanted was for the trackway to have been used as a cattle droveway. Towards the end of the last day of the project he found a single tip of a wing case of an *Aphodius* dung beetle and decided it was best to keep silent. He feared that one dung beetle would become a stampede of bovines.

A constant problem is what could be called the archaeoentomological 'Pavlovian response'. Like the ringing of bells causing salivation in dogs, the recovery of an 'indicator species' can bring on the worst bouts of 'textual' diarrhoea in archaeoentomologist. A few grain pests can lead to several pages of text outlining all that has ever been said concerning these species. A find of a single elytron of the elm bark beetle results in several paragraphs discussing the elm decline (regardless of the actual age of the deposit). Having just reread some of my own reports from London written early on in my career, I must admit to being very guilty of this crime. I ask the court to take into account my inexperience and youth at the time.

This over-emphasis can lead to some rather embarrassing moments at 'project meetings' held to discuss the results of an archaeological investigation. I have often been quite emphatic in my report; '*there are so many grain pests present that this deposit must have been composed, almost entirely, of grain'*, only to find subsequently that the plant macrofossil analyst has had little in the way of grain but lots of indicators for cess, meadow grasses, food waste and so on. The animal bone specialist also has lots of clear evidence for food waste. The intestinal parasites specialist also argues that the deposit is likely to be cess. The actuality is that the insect fauna is simply dominated by insects from a very small component of the wide range of rubbish that has gone into the deposit.

Of course the way around all of these problems is to avoid just relying on the 'indicator species' alone and use all the other available sources of information on the deposit. This is dealt with as a technique in Chapter 10 when the use of 'indicator groups' and 'indicator packages' (*sensu* Kenward and Hall 1997) is discussed. In the meantime, it is always wise to remember that a bit of common sense often goes a long way. Or as Confucius would have it: *The cautious seldom err*.

CHAPTER 7: SAXON AND NORMAN LONDON

EARLY AND MIDDLE SAXON LONDON

In AD 410, the city-based authorities of Britain received unwelcome news from Rome. Effectively it said that they had to look to their own defence against barbarian raids and that no fresh Roman troops were to be sent to help. This traditionally marked the start of the 'Dark Ages' and the decline of civic centres in Britain; London amongst them. There is of course debate about actually how marked this effect was. There is evidence to suggest that, at least in England, there was some continuation of Roman life and behaviour for the next century (Esmonde-Cleary 1989; Dark 1993, 2000).

There is a tendency nowadays to see the 'dark ages' as merely a rather 'light grey' period of continuance between Roman Britain and Anglo-Saxon England of the 8th and 9th centuries. However, this may not be the case for London (Cowie 2000). The origin of late Roman 'dark earth' in many archaeological sites in London, and what it represents in terms of how settlement changed in this period, is open to debate (Vince 1990). However, its presence has been used to suggest that many areas within the city walls may have reverted to farmland (Clark 1989; Vince 1990; Cowie 2008) or to low scale urban occupation, by the 4th century AD (Macphail *et al.* 2003; Bateman *et al.* 2008). Marsden and West (1992) actually see the decline starting in the second half of the 2nd century AD. Certainly there is a complete abandonment of the walled town by the early 5th century AD (Milne 1995; Perring 1991; Cowie and Harding 2000; Cowie 2008). Wheeler's (1934) 5th and 6th century AD 'sub-Roman slum' seems not to have existed and it is more likely that London was simply a ruined town inhabited only by occasional scavengers (Cowie 2000, 2008). Though several early Saxon rural sites, often containing sunken feature buildings (*grubenhaus*), do occur in the Greater London area they are usually of small scale, scattered and are limited to the brick earths and gravel terraces above the floodplain (Clark 1989; Cowie 2000; Cowie and Harding 2000; Cowie 2008).

By the 7th century, according to a range of historical documents, London appears to have again become a port and trade centre perhaps instigated by Ethelberg of Kent in the early 600s AD (Cowie 2000). Bede in 730 AD described London, or *Lundenwic*, in the early 7th century as *'an emporium for many nations who come to it by land and sea'* (Clark 1989; Vince 1990; Cowie 2000). There also are records of the founding of St. Paul's church in London in 604 AD (Clark 1989; Vince 1990; Cowie 2000). Nothing is known of this church, but it is generally assumed that it lay within the Roman walls, just to the south of the modern cathedral (Clark 1989; Vince 1990; Cowie 2000). Certainly *Lundenwic* appeared to be of some importance since it hosted a working gold coin mint after 640 AD (Clark 1989; Cowie 2000).

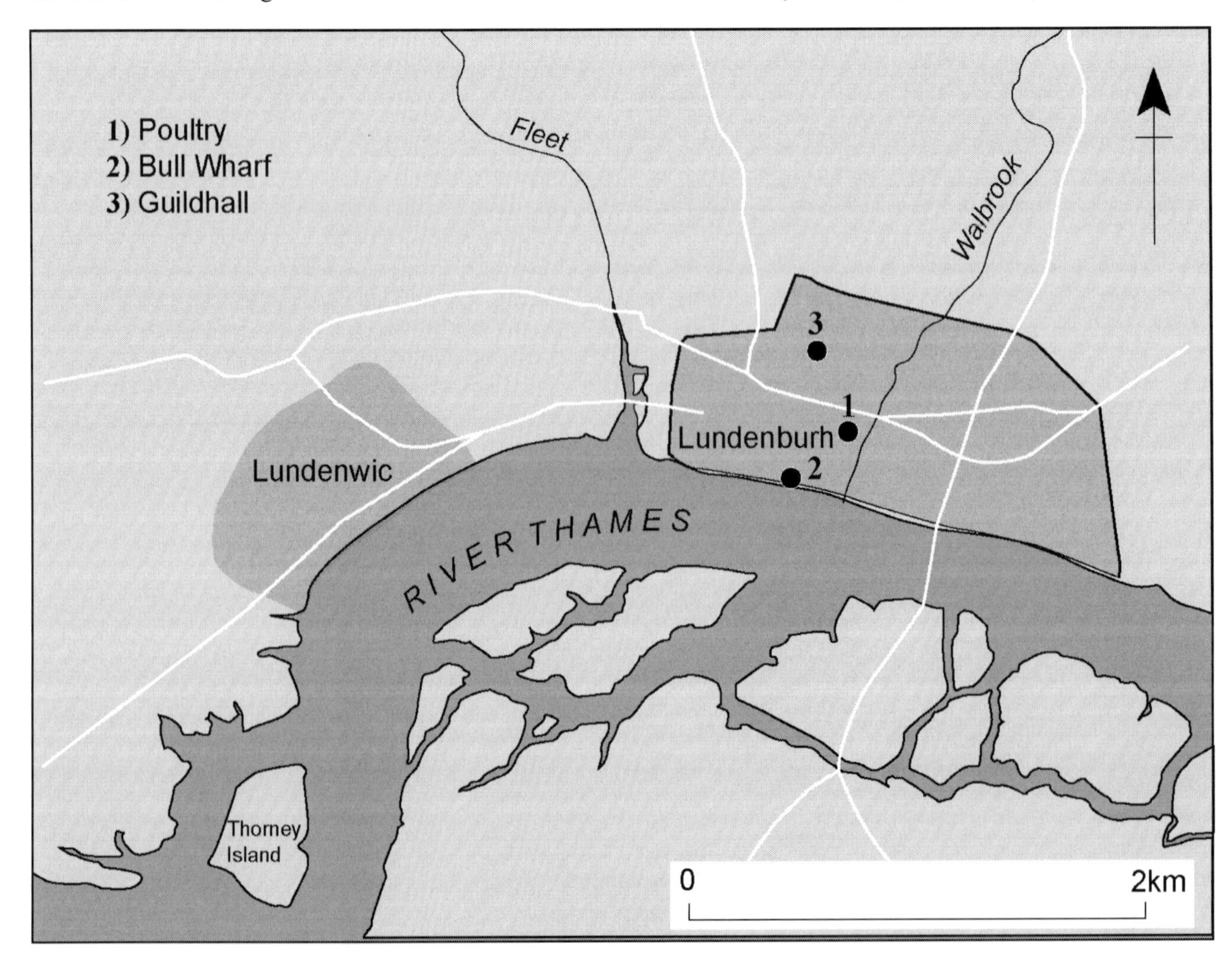

Figure 7.1. Sites from Saxon and Early Medieval London discussed in this Chapter

Until the 1970s the actual location of the site of *Lundenwic* remained a mystery. Wheeler (1934) presumed that it lay within the walled city itself, but the complete absence of any early and middle Saxon remains in excavations in the 'city' counts against this conclusion (Vince 1990). However, excavation in the area of the Strand and Covent Garden in the 1970s and 1980s produced evidence for dense Middle Saxon settlement (Whytehead 1988; Blackmoor *et al.* 1998; Cowie 2000; Cowie and Harding 2000; Malcolm, Bowsher and Cowie 2003), confirming the suggestion by Biddle (1984) that the Saxon settlement actually lay to the north in the modern Strand and Charring Cross areas. It is presumed that the settlement grew here because the Thames at this time was tidal up to this reach (Cowie 2000). Malcolm, Bowsher and Cowie (2003) suggest that in the early 600s this settlement may have been small and seasonal. The locations of the Middle and Late Saxon sites discussed in this Chapter are presented in Figure 7.1. The Opera House excavations (Malcolm, Bowsher and Cowie 2003) gave the impression that by the 700s AD Saxon London was a dense urban settlement, composed of substantial timber framed, wattle and daub buildings (Cowie 2000; Cowie and Harding 2000; Malcolm, Bowsher and Cowie 2003). The large amounts of imported pottery, and other goods, now recovered from this area of Middle Saxon London clearly show that it functioned as a major trade centre for East Anglia, with connections as far away as Norway and the Rhineland (Cowie 2000; Cowie and Harding 2000; Malcolm, Bowsher and Cowie 2003). Locally, the rural economy seems to have been concentrated on the production of wheat and barley for trade into town (Davis and de Moulins 1988; de Moulins and Davis 1989; Davis 2003). Similarly, meat production seems to have concentrated predominantly on cattle with small amounts of pig and sheep (Rackham 1994; Reilly 2003). Hodges (1982) and Cowie (2000; Cowie and Harding 2000) suggest that the 'rather dull' nature of the food supply is an indication that the town may have been provisioned directly from royal food rents rather than by trade throughout this period. *Lundenwic* seems to have had its heyday during the reign of both Mercian kings Aethelbard and Offa (716 – 796 AD) but to have declined again around 850 AD (Malcolm, Bowsher and Cowie 2003). Unfortunately *Ludenwic* has failed to produce, as yet, any waterlogged deposits and so we have no idea of Saxon London's insect fauna. Recovery of Saxon deposits suitable for archaeoentomological analysis, therefore, are of national importance.

LONDON AND THE VIKINGS

By the late 9th century AD the Greater London area began to fall victim to periodic Viking raids. This intensified by the mid 9th century, leading to major assaults on *Lundenwic* in 842, 851 and 871 (Vince 1990; Cowie 2000). The Anglo-Saxon chronicles also recount that a Viking army over-wintered in the area in 871 – 872 (Clark 1989). There is now growing evidence that one method to counter this increased threat after 851 was to re-occupy the walled Roman town (Cowie and Harding 2000). This is thought to have initially been a small-scale settlement between the river at Bull Wharf and Cheapside (Cowie 2000). Eventually this process culminates in Alfred's 're-establishment' of the town as a *burgh* (Cowie 2000; Cowie and Harding 2000) following its 'liberation' from the Vikings in 886. A grant of 888/ 889 makes it clear that the main Saxon settlement now was within the walls of the Roman city (Vince 1990). There is also considerable evidence that a new town plan, with new streets, was established during this time (Horsman *et al.* 1988; Vince 1990).

10TH CENTURY SAXON LONDON

In 1990 Alan Vince described the documentary sources dealing with London as *'surprisingly uninformative'* (Vince 1990, 26). It seems fair to describe our archaeological knowledge of the city up to the 1990s in a similar way (see Clark 2000). The lack of historical evidence has been attributed to the fact that London actually 'held out' against the Vikings throughout the 10th century and receives little attention in that arch 'Viking bashing text' the *Anglo-Saxon Chronicles* (Vince 1990). As Vince puts it *'no news is good news'* (Vince 1990, 26).

The late 1990s saw a number of crucial excavations at sites that date to this previously unexplored period. Dating deposits at this time is difficult, but two of the sites cluster between *circa* 950 – 1050 AD. Fortunately for me, both of these sites were waterlogged and throughout the late 1990s and early 2000s regular vanloads of samples were dispatched by the Museum of London up the M40 to me in Birmingham. It was only when I started to outline this book that the true significance of this work actually sank in. Bull Wharf (Thames Court) and Poultry give us our first real view of the nature of settlement, living conditions and urban life from this well-known historical period, with its tales of war, death, internecine feuding and general nastiness.

Bull Wharf

Excavations between 1990 and 1996 at Bull Wharf (now Thames Court) have given us an impression of how the Late Saxon 'trading shore' may have worked (Wroe-Brown 1988). The site lay to the south of Upper Thames Street and was the location of the Medieval dock of Queenhithe. The area was actually mentioned by name in the grant of 888/9 mentioned above. This allowed the Bishop of Worcester's men to trade in the area of the estate, but if they used the 'trading shore' they had to pay royal dues. The excavation revealed a sequence of foreshore structures, revetments and embankments that throughout the Late Saxon period encroached out into the river (Wroe-Brown 1988) to form a series of wharfs. Associated with these structures were a number of timber and earthen-floored buildings dating to the late 10th and early 11th centuries. The insect remains all come from this part of the site. Six samples (116, 117, 118, 121, 123 and 124) were studied from flooring deposits in Buildings 2 and 3. In fact, these are the first set of insect faunas that appear to come directly from flooring deposits studied from London to date. The floor deposits produced a relatively high proportion of the group of 'house fauna' species that Kenward (Hall and Kenward 1990; Kenward and Hall 1995) suggests lived in the litter that commonly built up on floors in buildings. These are species, such as *Xylodromus concinnus, Trogophloeus bilineatus, Cryptophagus* and *Atomaria* species, *Enicmus minutus* (group) and, perhaps, *Aglenus brunneus*. This is the group that Kenward initially labelled the 'house fauna' (Hall and Kenward 1990; Kenward and Hall 1995). More recently they have entered 'group A' in his re-analysis of the insect faunas from York (Carrott and Kenward 2001). The faunas suggest these floors built up slowly with relatively dry material being incorporated into the beaten earth. Many of these species may also have lived in nearby walling and roofing. As well as being used for thatch, there is the suggestion that reeds and other waterside vegetation may have been used for 'strewing' onto the floors (Buckland *et al.* 1974). This is suggested by a persistent presence of *Donacia* reed beetles and *Thyrogenes* and *Limnobaris pilistriatus* weevil, species associated with various water reeds.

So far so good, but many other species recovered suggest that far less pleasant material was dumped onto these floors or that conditions in the buildings became rather horrid at a later stage. Typical of these conditions are *Cercyon analis, C. atricapillus, Acritus nigricornis* and a range of *Trogophloeus*, *Oxytelus*, *Leptacinus*, *Neobisnus* and *Anthicus* species. This is, of course, Kenward's 'oxyteline group' (Kenward and Hall 1995) and its other associates that now form groups 'B' and 'C' (*sensu* Carrott and Kenward 2001). This grouping is thought to be typical of deposits consisting of rotten organic matter mixed with rather fluid mud. The range of flies suggests a similar environment. This includes small numbers of the common housefly, *Musca domestica,* and the stable fly, *Stomoxys calcitrans*. The *Sepsis* flies along with the Copromyzinae and *Thoracochaeta zosterae* also indicate rather fluid conditions.

There are two ways of looking at the presence of this mix of 'communities' in the same deposits, and within the same buildings. The first is that they represent a mixture of materials that co-existed during the occupation of the building, so we could think of patches of wet and dry flooring. Given the Thames-side nature of these buildings this is entirely possible. The second alternative is that the different materials/ faunas may represent occupation deposits on the one hand and/ or what happened to these after abandonment or burial.

What the Saxon London insects have not yet done is suggest the use of these buildings; whereas, elsewhere (e.g. Kenward's work at York, and a number of other sites – Hall and Kenward 1990; Kenward and Hall 1995, 1997; Hall and Kenward 2003) archaeoentomological data has clearly produced evidence for the specific use(s) of a building. At Anglo-Scandinavian York, Kenward (Kenward and Hall 1995, 725) was able to demonstrate both from the archaeological record, and from the insect faunas themselves, that the floor deposits were essentially occupation surfaces. He could then use the insects recovered to directly discuss living conditions and the range of activities that may have occurred in the buildings. This is clearly not the case with the floors at Bull Wharf where dumping or the re-deposition of material onto the floors seems to be the case. These faunas give the impression of it all being 'a bit of a mess'.

A similar impression is also gained from the insect faunas recovered from the other features at Bull Wharf. Many of these were interpreted as fill, midden or occupation deposits and the insect faunas have a strong resemblance to those studied from the floors of the buildings. Again, this suggests that the floor deposits from this site are later decay and/ or dumping of materials within these buildings. One possible exception to this is sample 734 from the midden associated with Building 16. The insect fauna is dominated by several hundred individuals of the small fly *Thoracochaeta zosterae* and by the latrine fly *Fannia scalaris*. Both of these species will be discussed in detail when we consider the fascinating world of cesspits in Chapter 11, but here they probably indicate that cess, or even the fill of a cesspit, was incorporated into this midden.

Unsurprisingly, many of the pit fills, particularly from the later phases of the site and from the 'barrel pits', indicate the inclusions of settlement waste and, to judge from the apparently flourishing faunas of flies, faecal material and 'cess'.

Poultry

There appears to be no substantial activity at the Poultry site from the late 3rd century AD up to the end of the 9th century. When the Late Saxon town developed the crossing of the Walbrook settlement moved to the north of the area and the ancient Roman roads were not strictly followed (Rowsome and Treviel 1998; Burch and Treviel 2011). In particular a new road was established that followed what now is the line of modern Cheapside. On the south side of this road a series of narrow wooden buildings with earthen floors and wooden or brickearth partitions were constructed towards the end of the 10th century (Rowsome and Treviel 1998; Burch and Treviel 2011) (see Figure 7.2.).

It also has been suggested that iron working was a substantial trade in this area of the town in this period (Rowsome and Treviel 1998; Burch and Treviel 2011). The insect faunas recovered from this period all come from a range of rectangular pits, some lined with wattle, from the areas between the two ranges of the buildings, or from directly behind the range running along the later route of Cheapside. Six samples of material from Poultry date to the period AD 970–1050 when these features were in use. These are samples 517, 693, 703, 740, and 798 and the ecological groupings (especially rt, rd and rf) from these features are shown in Figure 7.3, which indicates the dumping of settlement waste again. The range of plant material present also confirms this. This includes cereal bran, the seeds of corncockle, fruit stones and nutshells (Davis 2011). Along with the large amount of domestic animal and fish bone recovered, this suggests that food waste was common in these features. A large

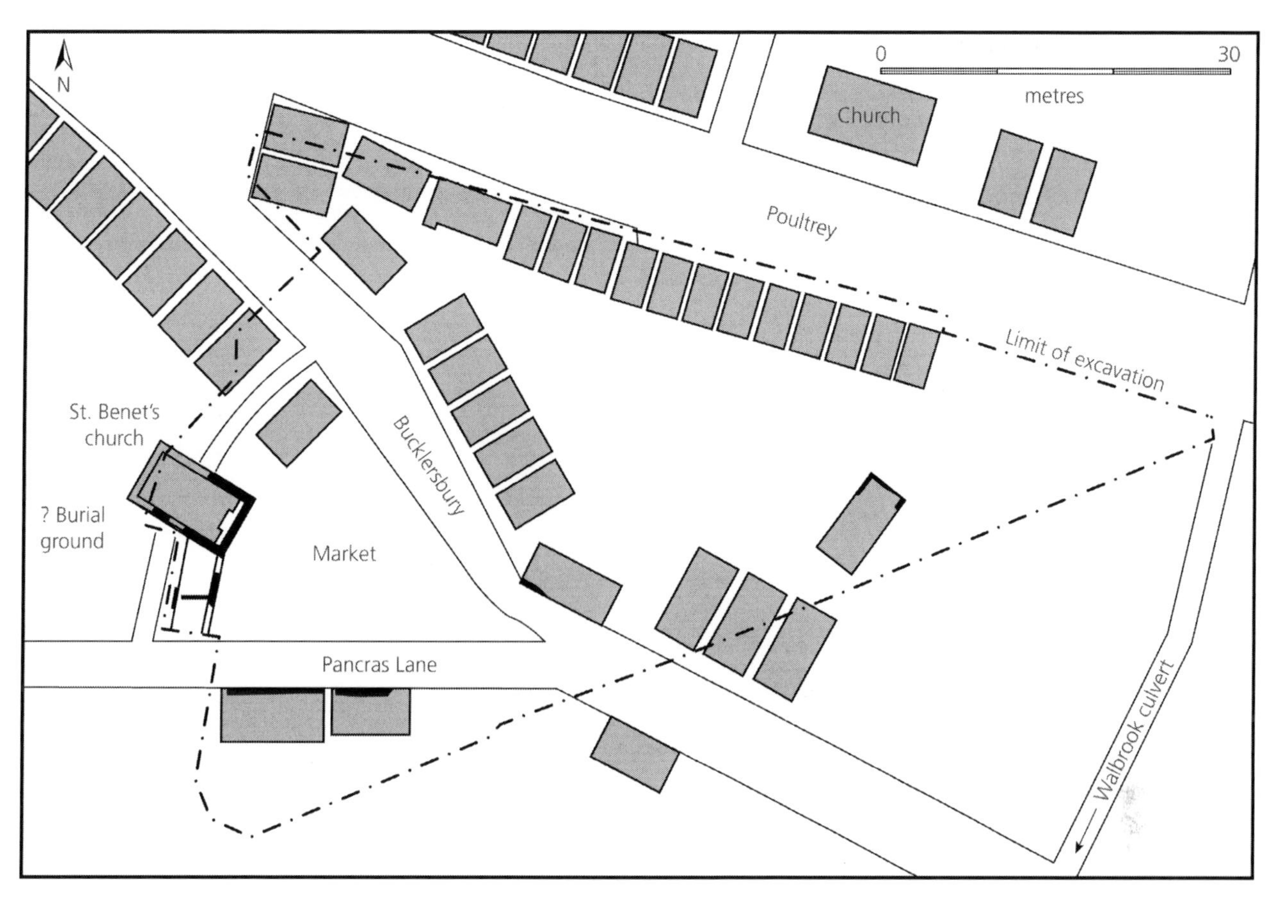

Figure 7.2. The Late Saxon and Norman features at Poultry (redrawn from Rowsome and Treviel 1998)

number of fly remains also suggest that considerable quantities of cess were present. This is clearly indicated by species such as *Sepsis, Telomerina ?flavipes* and *Thoracochaeta zosterae,* which all appear to be common inhabitants of cess and cesspits in the archaeological record (Belshaw 1989; Skidmore 1999). Several of the pits also contain the tube-shaped 'rat tails' of the larvae of the drone fly *Eristalis tenax*. This latter species is a rather specialised inhabitant of water containing high concentrations of faecal material and other foul matter. It floats just below the surface or on the bottom of shallow ponds and uses its 'rat tail' (or siphon) as a snorkel (Skidmore 1999; K.G.V. Smith 1973, 1989).

Large quantities of pea weevil, *Bruchus pisorum,* and some of the pests of stored grain such as, *Sitophilus granarius* 'the granary weevil', also were recovered in the pits. This may, of course indicate the deposition of spoilt agricultural products. However, it is more probable that these insects entered the deposit in cess after consumption of infested peas and grain. Cesspits and their context and contents will be discussed further in Chapter 11. However, it is clear that many of the pits at this site clearly fulfilled the same function and produced a similar insect fauna to those from the Anglo-Scandinavian site at Coppergate, York. Here the use of pits for disposal of rubbish and cess disposal seems to have been paramount as well (Kenward and Hall 1995, 748).

There are a few species in these pits that might suggest the environment in the yards behind the buildings. The 'ground beetles' recovered are all typical of rough disturbed turf and ground around human habitation (Lindroth 1974). Stinging nettle (the food plant of *Phyllobius argentatus* and *Cidnorhinus quadrimaculatus*), dock (the food plant of *Rhinocus pericarpius*) and plantain (the food plant of the *Gymnetron* species) were all present in the yards, suggesting rather weedy areas. Of course, it is possible that these species could have entered this deposit as part of the deposition of discarded flooring or stabling material containing hay (see Kenward and Hall 1997). However it seems likelier that this open pit functioned as a pitfall trap incorporating these species by accident. Similar sets of insects were commonly associated with the external yard deposits at Anglo-Scandinavian York where Kenward and Hall (1995, 738) also suggest the presence of weedy areas amongst the messy scatter.

The Guildhall: Saxon and Norman phases

Further north at the Guildhall site there is again evidence that the Roman buildings, in this case the Amphitheatre, fell out of use by the mid-4th century (Bateman 1997; Bateman *et al.* 2008). In the intervening years the area seems to have been used for dumping since there is 'dark earth' in the hollow left by the arena (Porter 1997; Bateman *et al.* 2008). The amphitheatre seems to have survived as a boggy hollow up to the early 11th century,

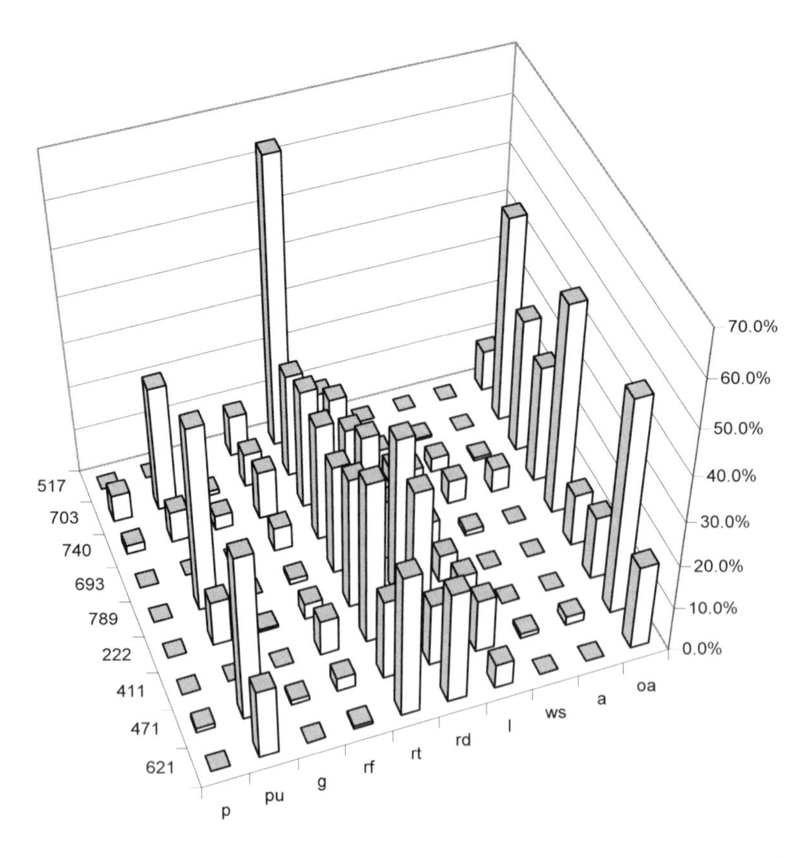

Figure 7.3. The ecological groupings of Coleoptera from the Late Saxon and Norman features at Poultry (ecological codes are outlined in Figure 1.2.)

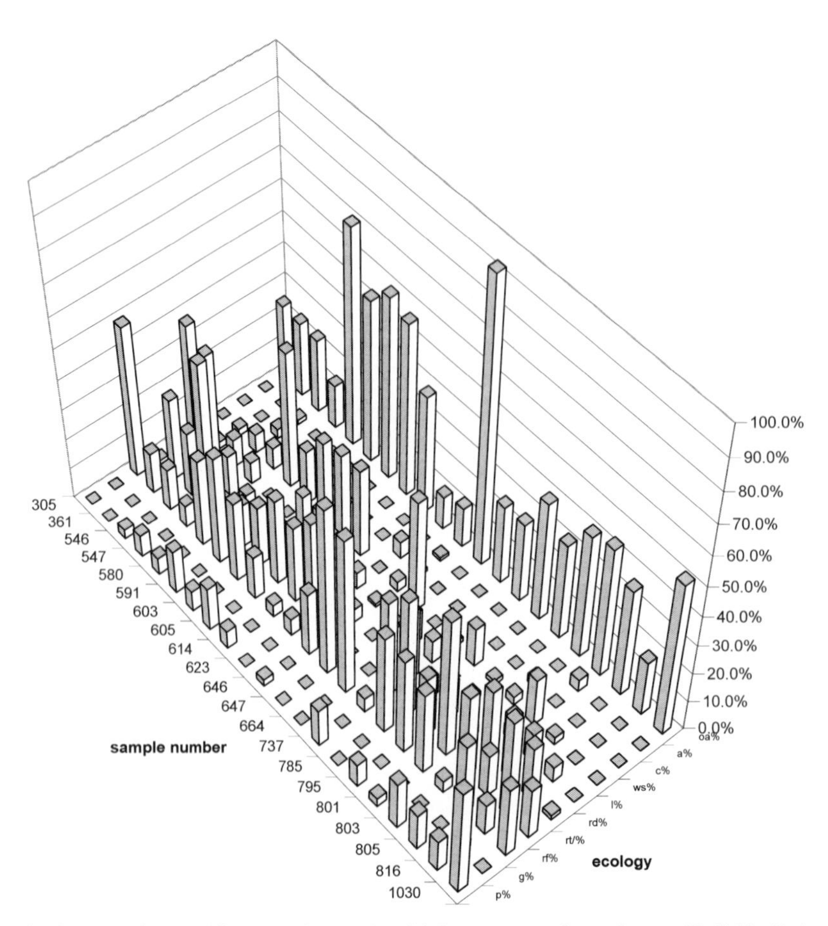

Figure 7.4 The ecological groupings of insects from the 11th century deposits at Guildhall (ecological codes are outlined in Figure 1.2.)

when the area was levelled and then used as part of the burial ground of the church of St Lawrence Jewry in the southeast corner of the site (Bowsher *et al.* 2008). The rest of the site seems to have been used as an area for cattle penning at this time (Porter 1997; Bowsher *et al.* 2008). After 1070 AD a range of domestic buildings with wooden and wattle walls and earth floors were established in this area along a narrow lane (see Figure 7.4) (Bowsher *et al.* 2008).

When study of the insect remains from Saxon and Norman 11th to 12th century deposits at the Guildhall was undertaken, it was hoped the faunas would provide clear interpretations of deposits and activities at the site similar to those seen at York. Certainly, given that we were to examine 32 samples from well-defined floors, dump deposits, external yard surfaces and pits, we thought we would gain a detailed story of the life and use of individual buildings. Unfortunately, this was not to be the case (Morris and Smith 2008).

The ecological groups for the insects from this period at the Guildhall are illustrated in Figure 7.4, from which it is clear that across the site there is a virtually ubiquitous fauna regardless of context type. In fact, boring is probably a better word. An M Phil. student undertook much of the identification of this material. Possibly understandably, the student has subsequently left the profession. I fear this may well have been the result of sheer boredom; six months looking at the same insects then another long period of time trying to say something exciting about them would be daunting to even the most experienced archaeoentomologist.

It is as though someone has taken several types of archaeological material 'blended' them together and then spread the result across the site. We see all of the 'usual suspects' both in terms of insect communities and materials that we have already discussed – but all at the same time and in the same sample. We seem to have a range of very fluid and wet, very decayed and 'blended' settlement waste. In essence, the material sampled and studied has not remained *in situ* after its primary formation but has been re-deposited several times. Both the insect faunas recovered, and the ubiquitous nature of the distribution of species between samples is very similar to those recovered from the early Medieval site at Highgate, Beverley where a similar interpretation was advanced (Hall and Kenward 1980). Many of the external layers from Anglo-Scandinavian York also exhibit the same phenomenon (Kenward and Hall 1995, 737).

In my final report to the excavator I said:

> *It would seem that much of the material from the buildings and open areas during both phases at the Guildhall contain mixed detritus and rubbish. This matter probably comes from a number of different sources and activities and therefore the material should be seen to represent an archaeological palimpsest.*

It is worth noting that a similar story seems to be true for the plant remains from these deposits (Giorgi 2007).

CONCLUSIONS

This chapter has clearly outlined the importance, at least in terms of date and context, of a set of insect faunas from the 9th – 11th century deposits in London. The main point to be grasped is that these are the only urban insect faunas from this period in London, and aside from York, and with some rather insignificant exceptions, the rest of the UK. They are clearly telling us much about life, living conditions and waste disposal in London at this time.

They also raise a number of questions, or rather interpretive issues that need to be discussed. In particular, much of the material examined is in fact very similar. Should we continue to examine 'run of the mill' faunas of these periods in such detail? Could I have spared my poor M. Phil. student her fate? This will be discussed in Chapter 8.

Nevertheless, we have also begun to see that with some deposits, particularly at Poultry and Bull Wharf, that the insects can tell very distinctive stories, particularly in terms of identifying the use and fills of pits. Chapter 10 will take us through the various arguments concerning how specific sets of insects can be used to identify deposits and human behaviour.

CHAPTER 8: WORKING AT THE SITE LEVEL

In the last Chapter we came across one of the most persistent problems of working on urban archaeoentomology. How should we deal with sites, such as Late Saxon/ Norman Guildhall, where a large number of insect faunas are recovered that are essentially the same? What should we do to reduce the duplication of effort involved? This is a problem that results from the fact that most archaeological deposits are 'palimpsests' in their nature. This is a concept borrowed from the examination of classical texts (Hurschmann 2009). Often the original document, to save the cost of using expensive parchment, will have been written over in several directions at different times. This means that the same document can carry several different messages that often run together in such a way that reading any of them can be very difficult. This seems the perfect analogy for what we see in the urban archaeological record. Often the material from individual contexts has been 'transformed' several times during the formation and use of a deposit. As with the example of the text, disentangling what has happened is one of the key problems of urban archaeology. It also is an intractable problem for many deposits examined in urban archaeoentomology where the message produced can, at times, be essentially 'smoothed' and unintelligible.

The real problem with examining large numbers of very similar urban deposits is that this work is usually done as paid consultancy. For the Guildhall site, processing of the samples took 10 days (4 samples per day), the identification of the insect faunas took 40 days (1 day per sample), and another 3 days were spent preparing talbes, summary statistics and writing the report. At the end of all this effort, one has to query whether it was worth the time, cost and effort of such an intensive study? Although informative, the results are not exactly earth shattering in their nature. We learned a lot about conditions in general but little in terms of individual archaeological deposits, the use of individual buildings or specific activities on site. In these days when the hunt for 'structured deposition' (e.g. Hill 1995; J. Thomas, 1999) is all, we often are left dealing with the most unstructured of deposits.

Now don't get me wrong. I am not suggesting that this work should not be done, as we see below there is still a great deal of useful information to be gained. Rather I argue that we need to routinely work in a different and more efficient way. The knee jerk reaction towards a full analysis, identification and quantification of all of the faunas present should perhaps be resisted. Of course the simple solution is ***assessment*** and limited quantification. This process will be discussed below in the conclusion of this chapter.

There also needs to be a general realisation that with urban sites we need to accept that the results are as significant at the site level as they are at the level of individual context. Indeed, I would go as far to say that for many archaeologists it is often the site level where much of their interest lies. Even at the Guildhall site, it was possible to draw up the standard list of 'interesting' and informative points that suggest the nature of life at this site. It also adds colour to our understanding of the past. In other words, what kind of questions can we commonly ask of even the most disturbed archaeological site?

WHAT IS THE NATURE OF DEPOSITION ON MY ARCHAEOLOGICAL SITE AS A WHOLE?

This is of course archaeoentomology's main area of strength and often is key to the understanding the formation of the archaeology of a site. At York, Hall and Kenward (Hall and Kenward 1990; Kenward and Hall 1995) were able to use the plant and insect remains recovered to clearly suggest that many of the deposits had not been re-worked and were essentially *in situ*. This meant that it was possible to discuss in detail how floor levels had formed, consider patterns of rubbish and cess disposal and possible building use and function.

This was clearly not the case at the Guildhall where deposits are essentially a 'palimpsest'. This should not be seen, however, as a negative result. For the Guildhall, the fact that the insect remains suggest major re-working and dumping of material, which clearly is of paramount interest to any archaeologist trying to interpret the stratigraphy of this site. Moreover, we should be asking what kind of activities could lead to a site such as this being 'turned over' to such an extent?

One approach, even on the most muddled of sites, is to look at how the ecological groupings of insects present vary between phases of activity. This was particularly successful at the Late Medieval site at Long Causeway, Peterborough. Here the main feature on site was the remains of the priory ditch (A.L. Jones 1996). Three main periods of activity were defined. During the 12th century the ditch seems to have been a dumping ground for a range of materials from domestic tenements, leatherworking shops and stonemasons that backed onto Long Causeway. This area of activity was over 15 metres away from the ditch at the time. The insect faunas were clearly from a very mixed set of materials. Like the Guildhall, the faunas were generally very similar in their 'mixed' nature. However, from the water beetles and reed beetles recovered, the early phases of the ditch had some areas of open water and aquatic vegetation. By the 13th century no water beetles were encountered suggesting that after this point the ditch was essentially 'filled to the brim', or at least above water level, with settlement rubbish. At some point in the 15th century the ditch was re-dug and replaced by a much less substantial feature. This smaller ditch produced a very different insect fauna. It contained a wide range of water beetles and species associated with aquatic plants and no indications for domestic or settlement rubbish.

So, even though we cannot use insects to look at the kind of detailed archaeological questions asked by Kenward at York when we have 'turned over' sites such as these, we can address issues of site formation and the general nature of the 'urban landscape'. In the case of

Peterborough it would have been fun to try to establish why we saw these changes, perhaps by using the historical record. Does it indicate that by the 15th century Peterborough had fundamentally changed the way that urban waste was removed, perhaps centralising it after this point? Indeed a similar response to rubbish and cess disposal is seen at this time in London and Winchester (Keene 1982). Or is it merely that by the 15th century the tenements had begun to back directly onto the priory ditch and none of the inhabitants wanted decaying rubbish immediately outside their window? Perhaps the authorities at the priory church began to object to the smell by this point?

Of course this sort of game, where general conditions and how they change are compared, can also be played between a number of different archaeological sites regardless of whether they have been turned over or not. One of the main points of this book is to directly compare my London results to Kenward's work from different periods at York (e.g. Hall *et al.* 1983, Hall and Kenward 1990; Kenward and Hall 1995), which will be discussed further in Chapter 11.

WHAT MATERIALS WERE BROUGHT ONTO MY ARCHAEOLOGICAL SITE?

This is a fairly straightforward question and only requires a look through the species list for taxa which could have arrived on site in fodder and building materials, or within materials that were required for trade, craft or manufacturing. A typical example of this would be the recovery of the various woodworms and longhorn beetles that indicate the presence of harvested and prepared timbers (Kenward and Hall 1995, 722). One notable example of this is from Roman deposits at Alcester, Warwickshire where sampling produced the remains of the longhorn beetle (*Hesperophanes fasciculatus*) which is found in southern Europe and not in Britain. This led Osborne (1971) to suggest that it must have been imported in prepared timber directly from the Mediterranean. Similarly, various species of beetle sometimes found in urban settlements are associated with heather (*Erica* spp./ *Calluna* spp.), rushes (*Carex* spp.), *Sphagnum* moss and acid bog conditions. A prime example of this is the small spiky weevil, *Micrelus ericae,* which feeds on heather. The presence of these species is commonly taken to be an indication of the importation of peat for construction or as a stabling material (Buckland *et al.* 1993; Kenward and Hall 1995, 724; Hall and Kenward 2003).

Many plant feeding species of beetle are associated with meadow hay and fodder. This is a resource that seems to have been frequently brought into town and appears to be a common component of a range of urban deposits (Kenward and Hall 1997).

Various individual species often are linked to particular crafts and their presence on site does at least indicate that this activity potentially occurred (Hall and Kenward 2003). A clear example of this is the presence of sheep keds; this parasitic fly has been 'fingered' as being particularly indicative of wool processing (Buckland and Perry 1989; Kenward and Hall 1995, 775; Hall and Kenward 2003, see also Chapter 5).

WHAT LIVESTOCK WAS KEPT AT MY ARCHAEOLOGICAL SITE?

This issue is again addressed by looking at the species list for the site to see which ectoparasites of domestic livestock are present. The use of lice in general was discussed in Chapter 5. A particularly good example comes from the early Historic Rath site at Deer Parks Farms, Northern Ireland where some floor samples contained abundant remains of the pig louse, *Haematopinus apri,* suggesting that pigs were kept (Kenward and Allison 1994). This site also contained numbers of cattle, sheep and goat lice (*Damalina bovis, D. ovis* and *D. caprae*) indicating that these – or their skins – were present in the buildings as well. This is in clear contrast to the situation at Anglo-Scandinavian York (Kenward and Hall 1995, 778) and the various sites from London examined here where only relatively small numbers of lice are recovered. This may suggest that live animals were not that common in urban settlements, or at least they may not have lived on site for long periods of time (Kenward and Hall 1995, 778).

WHAT FOOD WAS STORED OR CONSUMED AT MY ARCHAEOLOGICAL SITE?

This again usually breaks down to a discussion of grain beetles and how they end up on site, and the implications for grain storage. This was discussed in detail previously in Chapter 6.

WHAT WAS THE NATURE OF HEALTH AND HYGIENE AT MY ARCHAEOLOGICAL SITE?

This is a popular question to address with the insects from an archaeological site. We have already seen that the insects from urban sites often give an impression of very unpleasant and unsanitary conditions. Archaeoentomology can usefully provide the 'yuck' factor. For example the unpleasant state of the floors, yards and pathways seen in the 'Yorvik' reconstruction mainly results from the insect remains. Of course, as was pointed out in Chapter 1, we do need to worry about how typical these deposits would be in 'normal settlement' and to consider post-depositional factors. This is especially true of sites such as the Guildhall, Peterborough and Beverley where it is clear the site has been highly disturbed, turned over or covered in dumped deposits. However, there does seem to be a general pattern here, and we should perhaps have more faith that these do represent real 'living and lived on' deposits. Certainly, Kenward (Kenward and Hall 1995, 727) found that the insect fauna from the well-preserved floor deposits at Coppergate, York contained many components of insect fauna that are found in these very mixed urban deposits. This might suggest that they are, to some extent, actually indicating 'average' conditions on site.

Of course, unpleasant is a relative term. Dealing specifically with the nature of the floors at York Kenward and Hall remark:

> *What the average householder of the late 20th century would perceive as filth may well have been regarded as a cosy litter in Anglo-Scandinavian York. Even the fleas may well have been regarded with affection* (Kenward and Hall 1995, 731).

The same is probably true in terms of how the general urban environment may have been conceived. The noise, filth, flies, fleas and the need to '*garde l'eau*' may have been one of the attractions of the urban scene.

One common thing said about urban settlement in the past is that living in it was bad for your health (Brothwell 1994; Manchester 1992). Insects can help promote this argument. Even when analysis is limited to the 'site level' one predominant characteristic of urban insect faunas is they all contain large numbers of flies. Amongst the most common are the housefly (*Musca domestica*), the stable fly (*Stomoxys calcitrans*) and the lesser housefly (*Muscina stabulans*). These species have all been implicated as vectors in a range of diseases. Usually, this consists of various forms of bacterial diarrhoea and fevers probably brought into town by adult flies that alighted on food (K.G.V. Smith 1973). This is not a small matter, the 'shits and the shakes' may be solved quickly today with anti-diarrhoeal drugs, aspirin and antibiotics but could kill quickly in the past, particularly the young and the old, though higher immunity in the past may have been relevant here (Kenward *pers. com.*). Kenward and Large (1998) suggest that the smaller flies, such as *Thoracochaeta zosterae*, and adult beetles may well perform the same role. The real villain of the piece, of course, is the housefly, which is also known as 'the dark destroyer', and has been implicated in the occurrence of viral poliomyelitis, dysentery, typhoid, tuberculosis, leprosy, plague, amoebic dysentery, tapeworms and intestinal parasites (K.G.V. Smith 1973).

Other vectors for disease that are always mentioned are fleas and lice. The usual suspects are the human flea (*Pulex irritans*), the head louse (*Pediculus humanus capitis*), the body louse (*Pediculus humanus humanus)*, and the pubic louse (*Pthirus pubis*). The louse is held to be the vector of relapsing fever, endemic typhus, trench fever and other conditions. Large louse populations on a single individual can lead to a general malaise – the advent of our common expression 'feeling lousy'. It has also been 'fingered' in the past as the villain behind the Athenian plague of 430 – 437 BC (McArthur 1958; Wylie and Stubbs 1983). *Pulex irritans* (the human flea) also has been implicated in various plagues. However, I feel that these creatures often are unjustly maligned. In the normal course of events they do not represent a great threat to human health. Moreover, the medical diagnosis of plagues in the past is notoriously difficult (Holladay and Poole 1979). In fact, fleas and lice carry a limited range of diseases; after all, who wishes to kill off one's host?

My wife has had personal experience of human fleas and did not find them that difficult to live with. The first year she worked on the Amarna project in Egypt the dig team found that they had small bites around their waists and ankles. After a mini-beast safari they collected some of the culprits for me. They naively believed these were merely dog and cat fleas. Well, they were not. They were human fleas. The next year she took out a cream that contained eucalyptus and cedar oils. This seemed to reduce the number of bites that occurred and amount of wildlife running around in her clothes. In fact this was such a success that the rest of the crew 'borrowed' the supply. I got a panicked phone call from Egypt asking me to dispatch more of the cream ASAP. So I went to DHL. 'When will this get there?' I asked. 'By 12.00 the next day, guaranteed' came the reply. What they did not know was that it had to go to Cairo, get through Egyptian customs and then meet a military convoy and travel 200 km across the western desert to get to the isolated site. I went away chuckling at DHL's over-optimism.

Behold, at 11.30 the next day, on the distant horizon at Amarna, a cloud of dust from a small convoy appears through the shimmering light. Eventually a tank, two armoured vehicles and a truck full of troops surrounding a tiny DHL van came into view. This all pulls into the dig compound and the vans doors are thrown open to reveal a small box of herbal cream. As with most Chinese whispers, the small bag of anti-flea cream had become 'urgently needed medication' by the time it arrived at Amarna. Nobody had the heart to tell the driver otherwise.

Returning to our theme, most urban deposits appear to carry very large numbers of intestinal parasites such as the whipworms (*Trichuris*) and the maw worm (*Ascaris*) (A.K.G. Jones 1983, 1985). This suggests that throughout the period of urban settlement populations carried a relatively large 'load' of intestinal parasites. However, the implications of the presence of such large populations of intestinal parasites for human health are not clear. Once again these parasites often do not act as a vector for disease or directly cause sickness. Again, why kill off one's host? At this point I must admit to having a personal interest in this. Most schools suffer infestations of head lice. Mine went in for pinworm outbreaks. None of us 'dirty little boys' felt the least affect except for the fact we were force-fed a worm powder that contained a raspberry flavouring that did not really hide the disgusting taste. To this day I cannot taste a raspberry milkshake without gagging (one of my colleagues here at Birmingham went a funny colour when I related this story in the coffee room. Obviously, this part of her childhood not only came back to her at speed but also with an explanation she did not really wish to hear).

So if we cannot use the presence of flies and the various parasites as *a priori* indicators for the presence of a range of 'urban diseases', what could we use? In fact this a very difficult issue to address. Collections of contemporary skeletal remains for many of the Roman and Saxon/ Anglo-Scandinavian periods are relatively rare (Brothwell 1982; Redfern and Roberts 2005). In addition, many of the diseases we have discussed here either leave no diagnostic marks on human bone or kill so quickly that there is no time for such a development (Brothwell 1982; Redfern and Roberts 2005). One other possible approach might be to look at the plant macrofossil results from these sites and see if any of the species recovered could have been used as herbal medicine to help alleviate such illnesses (Kenward and Hall 1995). This, however, gets us into the problem of interpreting ancient herbals and records of diseases (Sally Crawford *pers. com.*). There are many other uses that these plants could have been put to, in addition to that of medical herb (e.g. G. Jones *et al.* 1991; Kenward and Hall 1995, 670).

Here we hit the same old problem seen throughout this book. We are dealing with potentialities and plausibilities rather than certainty. I now find myself looking back at some of the 'health and hygiene' sections of academic site reports that I have written and cringe. The temptation to fall into a 'Pavlovian response' in which the presence of flies and fleas automatically mean that disease is present (possibly rampant) on site is one that we should all struggle to resist.

HOW BIG WAS MY ARCHAEOLOGICAL SETTLEMENT? DID IT TRADE WIDELY OR WAS IT ISOLATED? HOW CONTINUOUS WAS THE SETTLEMENT?

Kenward (1997) has recently suggested that the nature of the synanthropic insect fauna present at an archaeological site may give us an insight into the longevity and isolation of settlement. This is based on an application of one of the main tenets of modern ecology: the concept of 'island biogeography' (MacArthur and Wilson 1967). This is a principle that has been advanced to explain why the flora and fauna of offshore islands appears to be reduced the further away from the mainland you travel. Three factors are commonly considered:

1) the size of the island – small islands tend to contain less ecological niches and so smaller faunas
2) The distance from the source or mainland – the more isolated the island the less chance that species will reach it
3) The length of time in which the island is available to be colonised – obviously the longer the period the more chance that migrants will reach the island

Kenward (1997, 136) suggests that settlements in essence act as 'islands' of habitat for many decomposer and synanthropic species of insect, with these accruing on site from distant sources or 'urban continents'. He (Kenward 1997, 137) postulates that the size, diversity and membership of the decomposer and synanthropic insect fauna at an archaeological site will depend on:

1) How near it was to existing settlements (the nearer to a nearby settlement the more likely that the species could 'migrate' between them)
2) The degree of trade between the sites (this would hasten the spread of species since they could be carried in trade goods – the grain pests come to mind)
3) How long the site is occupied and its size (the more stable and larger a settlement the more likely that it will develop a high level of ecological complexity and the synanthropic species will become breeding populations)

Kenward (1997) suggested that we should be able to use urban insect faunas to assess trade contacts and the longevity and intensity of urban settlement. The key may well be the presence of species that are particularly strongly associated with human activity. This includes many of the 'house fauna' along with a range of extreme specialists such as the tenebrionids (darkling beetles) and the demestids (hide and carpet beetles), which are regarded as being strongly dependant on human activity.

This is a very good 'thought model' that Kenward backs up with a number of archaeological examples from a range of large and small archaeological sites from a number of differing periods which appear, in general, to agree with this proposition. One contrast given is between the isolated Rath at Deer Park Farms, Co. Antrim, N. Ireland and 9th to 10th century York (Kenward and Allison 1994; Kenward 1997). However, one site that I worked on in the 1990s has led me to question how useful this model is. I am not so sure that the process involved in the colonisation of synanthropic species is quite as slow or sensitive to contact as Kenward suggests.

The Iron Age buildings from the foreshore at Goldcliff, Gwent are extremely isolated. At the time of their use they appear to have been located some distance away from the 'landward' farms and villages on the hill slopes overlooking the vast expanse of the moorland and peat bog that made up the Iron Age Gwent Levels (Bell *et al.* 2000). These small buildings were used for grazing cattle on the foreshore and salt marsh that dominated the area. This activity also appears to have been seasonal with the floor in one of the buildings clearly showing alternating layers of occupation and estuarine clay (Bell *et al.* 2000). Despite this isolation, the limited nature of the settlement, and the brevity of both use and occupation, the insect fauna recovered is dominated by synanthropic species including many that Kenward sees as particularly significant. It may be that a single summer season is all that is needed for most of the synanthropes to establish a large breeding population once they arrive. One alternative is to suggest that large amounts of fodder, containing the synanthropic insect species, were carried out to site across the bogland. However, this does not appear to be the answer. There is no indication for this amongst the plant remains recovered; instead, the floor layers are dominated by reed (*Phragmites* sp.) and sedges

(*Carex* spp.), both of which would have been in plentiful supply locally (Casteldine 2000).

HOW TO PROCEED WITH DISTURBED SITES AND ISOLATED DEPOSITS?

From the above it should be clear that we could work with disturbed deposits and 'turned over sites' and still produce a range of informative results. We do lose the detail and we certainly do not have great 'context resolution', but in general 'randomised sites' such as the Guildhall do produce a worthwhile study. What we need to ensure is that these important results do not come at the expense of the project budget.

Quite often a large archaeological site will produce a very limited number of contexts that are waterlogged. Should these all be studied? Are they representative? The answer, of course, is to insist that all archaeological sites should have an assessment stage built into the analysis. This is of course was a requirement of the English Heritage 'Management of Archaeological Project 2' document and has been incorporated into its successor 'Management of Research Projects in the Historic Environment (MoRPHE)', but this stage is often overlooked or neglected.

Assessment of the environmental record from archaeological sites, and insects in particular, has been widely discussed (Kenward *et al.* 1985; Kenward 1992; Association for Environmental Archaeology 1995; English Heritage 2002). In terms of the various problems raised by this chapter an assessment of the insect faunas present in a urban site's samples would enable us to identify if deposits are mixed or not, and to see if they have an interpretative role to play at the level of the individual context or 'merely' at that of the site as a whole.

Initially, it was suggested that merely one litre of material from the sample should be processed and the insects recovered 'scanned' for a few minutes. If it looks well-persevered, interpretable and of interest more material should be processed at a later stage (Kenward *et al.* 1985). This still is a technique often routinely employed, particularly for evaluation excavations and large urban sites.

There is a problem here. I often have found that the insect fauna from one litre does not really give a clear impression of the 'true' nature of the insect fauna that could be recovered. In addition, there often is a logistical problem, particularly if the archaeoentomologist changes between the assessment phase and the main post-excavation project. You can easily end up with various sets of materials in different places, some of which may even get lost along the way. I recently have had to carry out final and full analysis based on the insect remains from the litre of sample examined for assessment, because the general biological sample of 10 litres of unprocessed sediment was inadvertently discarded. Not an ideal state of affairs. Moreover, the unprocessed soil samples often degrade in the period between the assessment and the main post-excavation phase, which can be several years, even decades. I am afraid that these days I tend to recommend that the entire sample taken for archaeoentomology is processed and the insects from it are sorted and stored in ethanol or another preservative at the start of the assessment stage of the project. This does make the assessment phase more expensive but it means that the assessment gives a clearer picture of the nature of the insect faunas present and how to proceed with them. In addition, you end up with the whole fauna in one place and, because it is in the ethanol, it is stable for the long-term, and the insect remains will be present when you need them in several years time for the post-excavation analysis, even if that is several years later.

So if we cannot save time and money at this stage when can we? A detailed assessment does enable us to identify which samples are poor in preservation and content, which are 'mixed' and which are going to give us good context-based information. It essentially enables us to zero in on which faunas from an archaeological site are going to give us the best information and will produce the most useful interpretations. Sincerely, this does, in itself, save money in the long term.

But what to do about sites such as the Guildhall where we face many insect faunas that are essentially the same? Well, first perhaps I should have spared both the student and myself some pain by undertaking an assessment before we reached full post-excavation analysis. However, in this case this was not really possible. The preservation and quantity of the insect remains had been judged by the archaeobotonists on the project during their assessment. This is a cheap solution and commendable, often resorted to by commercial field units, but it does have a major weakness. It is difficult for archaeobotanists to assess the *interpretative* quality of the insects present and the extent to which they are going to help us understand the nature of the site as whole and individual deposits in particular. **The 'scanning' of insect faunas during assessment needs to be carried out by a trained archaeoentomologist**. Additionally, I would argue that it needs more skill, training and experience to assess material than to do the full analysis.

In terms of saving time during the main phase of post-excavation Kenward (Kenward *et al.* 1986; Kenward 1992; Kenward and Hall 1995) suggests that we should put into place a sliding scale of analysis depending upon circumstances and the quality of the insect fauna suggested during the assessment. With deposits such as the Guildhall I tend to now suggest that a 'semi-quantitative scan' is carried out. This is a 15–20 minute scan of the material, often in the ethanol rather than carefully laid out on filter paper. All taxa seen are noted and identified where possible. Unknown sclerites (beetle heads, wing cases, etc…) are pursued only if they are thought to have significance. A five point 'star scheme' is used to quantify the material (usually scored as 1, 5, 10, 20 and >20). This is a lot faster than the full analysis of a single sample, which usually takes a day. In general semi-quantative scans can result in the appraisal of between

12–18 samples per day but, despite this speed, the process still producing results that are of good quality. Moreover, this semi-quantitative approximation of number of individuals can be used in statistics (Kenward 1992, 84). I have used this technique very successfully. It was used at the Peterborough site discussed above and the Bull Wharf and Preacher's Court sites from London. During the writing of this book I returned to these faunas, fully identifying and quantifying all insect remains. There actually is a remarkably high degree of correspondence between the taxa noted and the number of individuals suggested between the 'semi-quantitative scan' I initially carried out and the full analysis I performed years later for this publication (a conclusion which has also been reached independently by Kenward (1992) as well). I really must urge excavators and archaeoentomologist to consider undertaking this 'shortened' form of analysis more often. Larger numbers of samples and contexts can be examined and those 'special deposits' that merit fuller attention can be identified for full analysis.

CHAPTER 9: MEDIEVAL ARCHAEOLOGY AND LONDON'S INSECTS

In previous chapters, I have tended to start by outlining the archaeology and history of a specific period in some detail. I also have attempted to flag up major research issues for the periods. I am afraid that for the high and late Medieval I am not going to go into such depth. One clear problem with trying to attempt to bring together a potted history of London during this period is the wealth of information available. The problem can be clearly seen in the way Medieval studies in London have become highly specialised. A reflection of this can be seen in the structure of the *London Underground* book (Haynes *et al.* 2000). This deliberately set out to be a complete survey of how the archaeology of London changed and advanced in the thirty years before its publication. Most of the periods discussed are contained in a single chapter. However, the survey of Medieval London is divided into a number of separate discussions of topography, buildings and defence (Schofield 2000), pottery and trade (Vince 2000), burial (Harding 2000) and the archaeology of the playhouses (Blatherwick 2000). Despite the comprehensiveness of this survey there is still a feeling that this is an incomplete sketch. This is not a criticism of the authors but rather an indication of the potential trouble out there for the 'summariser'.

In terms of insect remains studied from Medieval London, we only actually have a handful of sites. Furthermore these all come from one particular site type, the great religious houses of Medieval London. I do not think there was a deliberate attempt by MoLAS to concentrate on this type of settlement, but by happenstance that is what we have (Sidell 2000). The locations of the Medieval sites discussed here are illustrated in Figure 9.1. Notably, none are located within the Roman or Saxon walled city.

MERTON PRIORY

At Merton Priory, I have analysed insect faunas from a number of Medieval river channel deposits near to the Augustine Monastic house (Miller *et al.* 2006; Smith 2006). However, these are not from urban settlement, as such, and will not be discussed any further in this book.

THE AUGUSTINE PRIORY AND HOSPITAL OF ST. MARY SPITAL, BISHOPSGATE

The insect faunas recovered from St. Mary Spital are discussed in Smith 1997b. This large Augustine Priory and Hospital was located outside of the Medieval walled city, north of the gateway at Bishopsgate. Walter Brunis, a prominent Londoner and possibly archdeacon, is generally believed to have founded the priory and

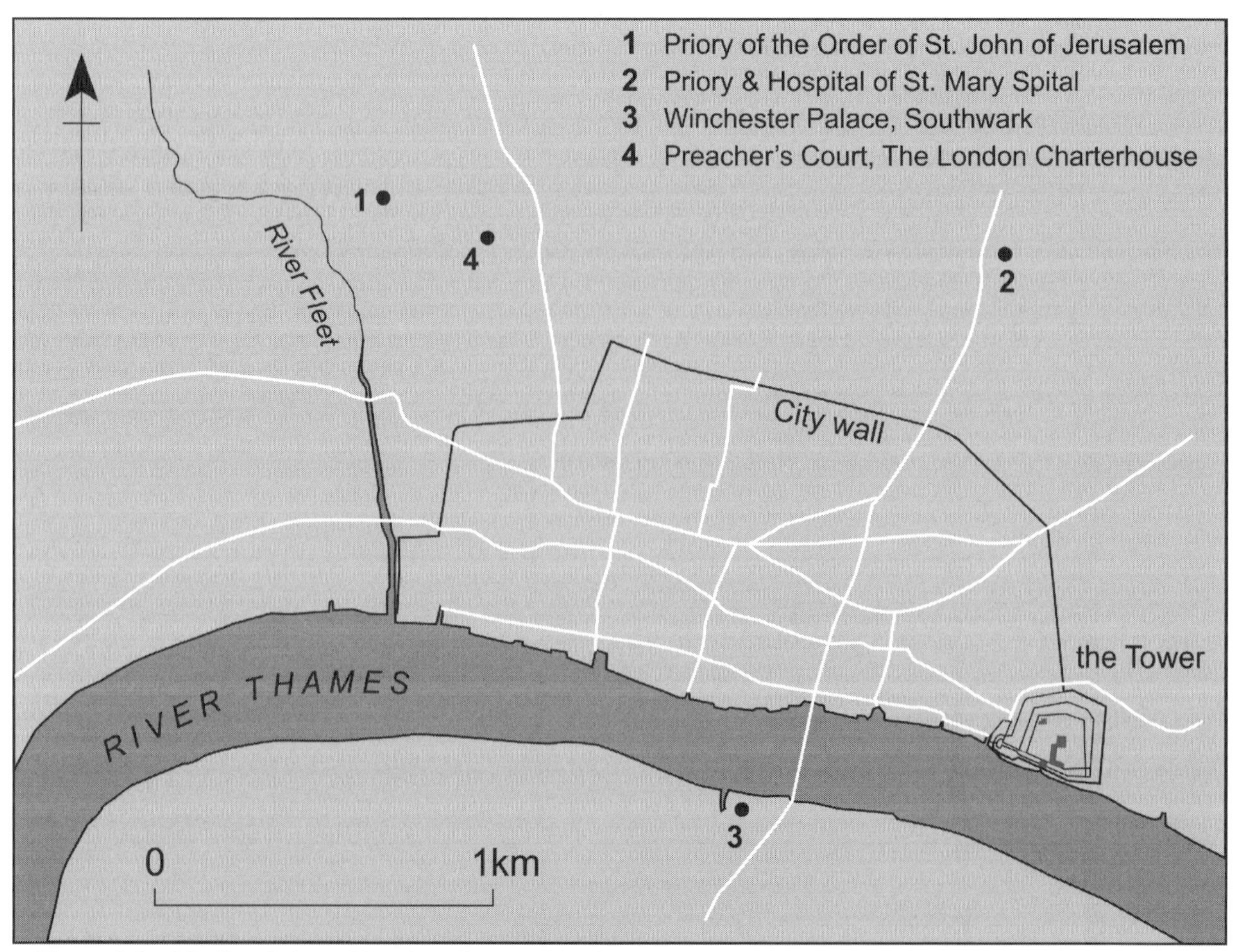

Figure 9.1 Medieval sites of London discussed in the text

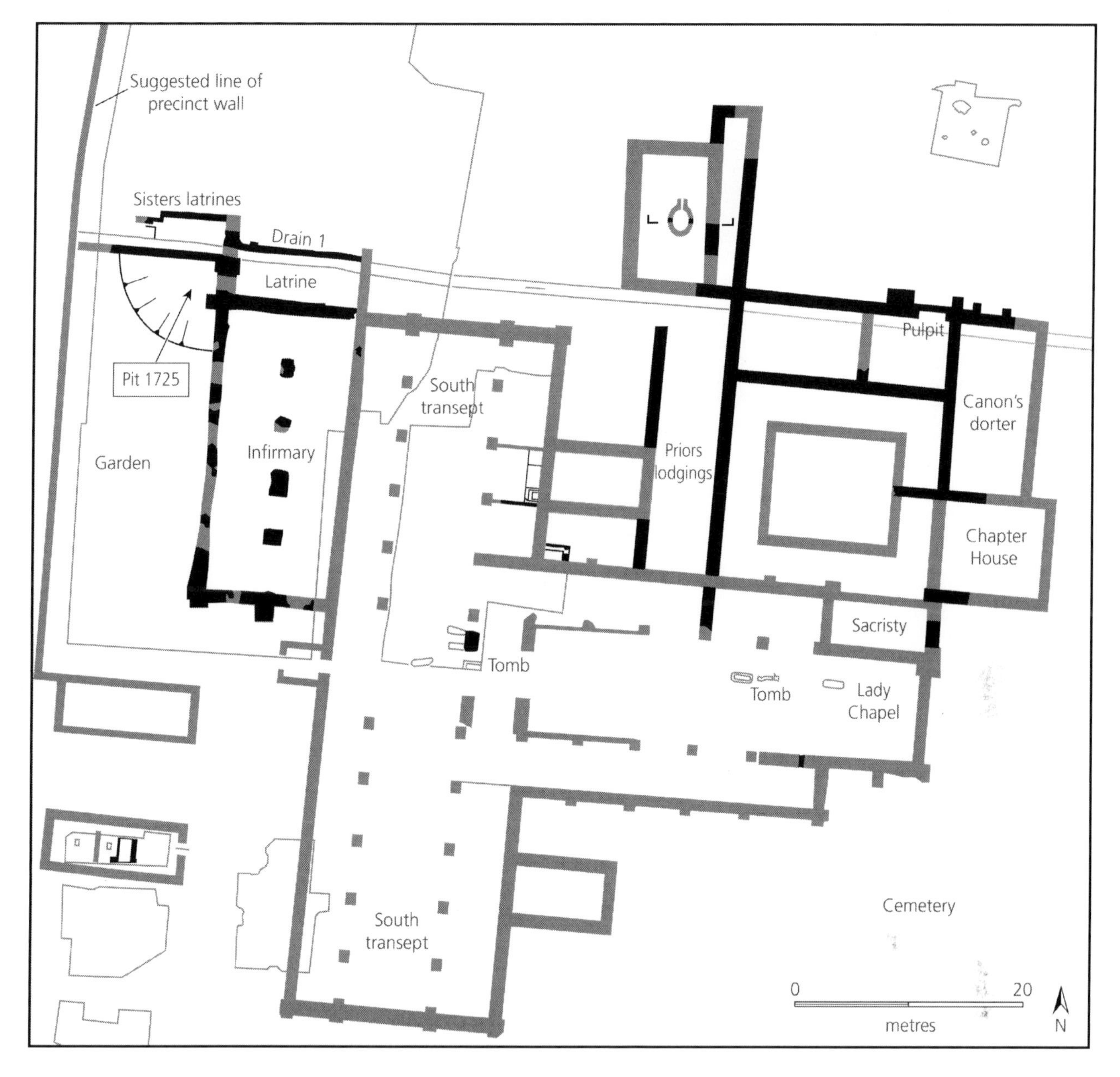

Figure 9.2. The location of the 13th and 14th century deposits at St. Mary Spital (redrawn from Thomas *et al.* 1997)

hospital in 1197 (C. Thomas *et al.* 1997). Alternatively, if it was not founded in 1197 then 1212 seems the next most likely date (C. Thomas *et. al.* 1997, 19).

Between 1280 and 1320, the Priory, its church and hospital had considerably expanded (C. Thomas *et al.* 1997, 43). The insect faunas analysed date to this period and come from a large pit (Feature 1725: samples 568 and 1286) that lay in the corner of the garden to the west of the new infirmary (Figure 9.2.). A sluice that led down from the latrines of the Sisters' quarters and the Infirmary drained into this pit. The pit itself contained a highly organic black deposit that included wood bark, twigs, pottery, shoes, wooden vessels and platters along with evidence for food waste (Thomas *et al.* 1997). The evidence for food waste comes from the plant macrofossil remains which include seeds of grape (*Vitis vinifera* L.), plum (*Prunus domestica* L.), cherry (*Prunus avium* L.), strawberry (*Fragaria x ananassa* (Duchesne) Duchesne), cereals, fennel (*Foeniculum vulgare* Mill.) and both white and black mustard (*Sinapis alba* L. and *Brassica nigra* (L.) W.D.J. Koch) (Davis 1997). Typically this type of flora is identified as indicating cesspit fills (Davis 1997). The insect fauna from this feature and its interpretation will be discussed below when we look in detail at life in the cesspit.

Later deposits from this site date to after the dissolution of the monasteries by Henry the VIII in 1538. Post-dissolution the property was leased to a series of courtiers who converted it into housing or commercial workshops. The insect faunas examined from this period are from a series of large pits excavated into a 'garden' to the north of the building that had initially been the hospital (Figure 9.3).

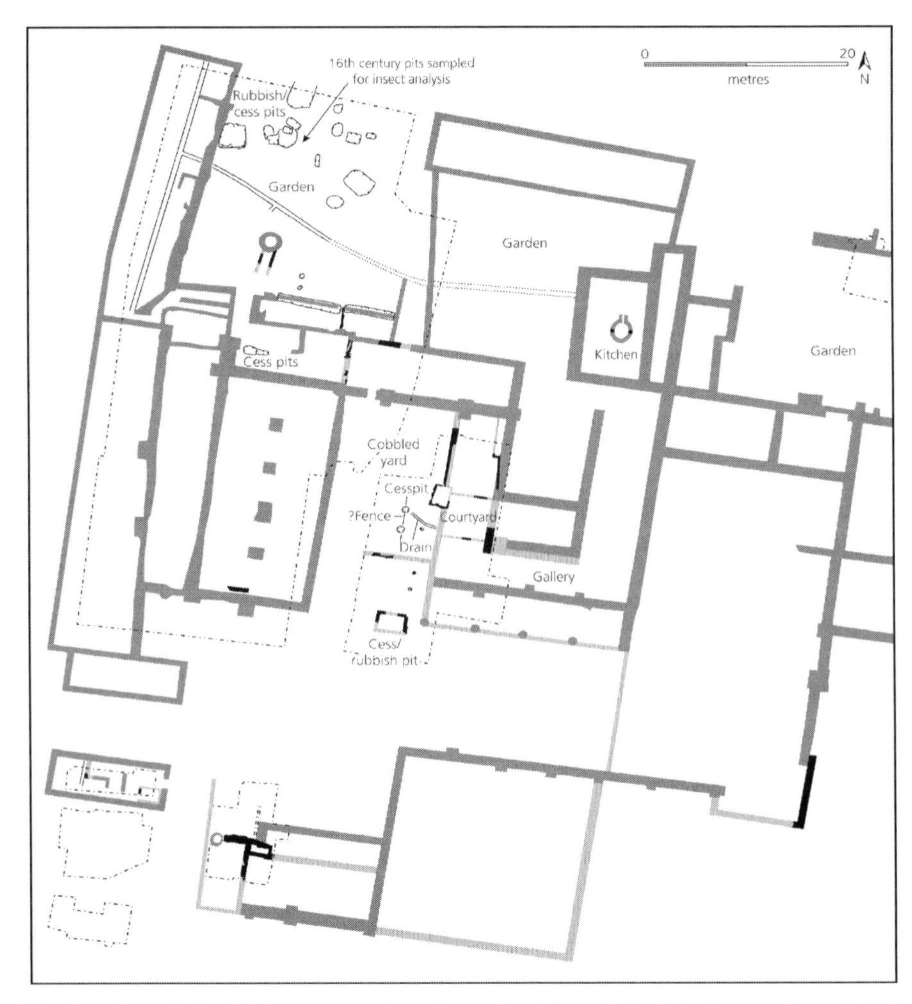

Figure 9.3. The location of the 16th century deposits at St. Mary Spital (redrawn from C. Thomas *et al.* 1997).

Three pits (171, 268 and 270) all produced large insect faunas. These were dominated by species associated with human habitation, and all were members of Kenward's 'house fauna'. However in pit 270 the fauna consists of the members of 'the house fauna' that come from the drier end of the spectrum (Kenward 1982). This includes lathridiid 'plaster beetles', ptinid 'spider beetles', *Xylodromus concinnus, Typhaea stercorea* and cryptophagids. This probably suggests that quantities of hay waste or other dry plant material had entered this deposit. The pottery, animal bone and plant remains indicate dumping of domestic rubbish also contributed to this deposit. This includes various food plants such as plum, fennel, mustards, poppy seed (*Papaver somniferum* L.), cucumber (*Cucumis sativus* L.) and quince (*Cydonia oblonga* Mill.) (Davis 1997). The presence of flax (*L. usitatissimum* L.), hop (*Humulus lupulus* L.) and hemp (*Cannabis sativa* L.) remains suggest that rubbish from industry or craft may also have been incorporated into this feature. Likewise, clippings of holly (*Ilex aquifolium* L.) and privet (probably *Ligustrum ovalifolium* Hassk.) suggest that garden waste also entered this deposit (Davis 1997). Faeces also appear to have been a common constituent since both pits 171 and 270 contained ova of intestinal parasites (de Rouffingnac 1997).

The samples from pit 268 (268, 187) tell another story. They are dominated by two grain pests (accounting for two thirds of the total fauna). These are the 'granary weevil' and the 'saw toothed grain beetle'. This suggests that the same problems that plagued grain storage in the Romano-British period in London also periodically occurred in later periods. This pit also contained the skeletons of over 30 gulls (Pipe 1997) and a range of waterside plants such as sedge (*Carex* spp.) and spike rush (*Eleocharis palustris* L.) (Davis 1997). This suggests that, in addition to dumped grain, material and rubbish from a range of craft and settlement activities were deposited into this pit. The impression gained is that these pits were at the back of a series of busy tenancies and accumulating waste from the wide range of activities that occurred within them.

PRIORY OF THE ORDER OF THE HOSPITAL OF ST JOHN JERUSALEM, CLERKENWELL

This large religious house was about 400 meters to the northwest of the city wall near Newgate. It was also to the east of the river Fleet and on the route running north to St. Albans. The Priory was the provincial headquarters of the military religious order of the Knights Hospitallers (Sloane and Malcolm 2004). Jordan de Bricet and his wife Muriel founded the priory in 1144. The site

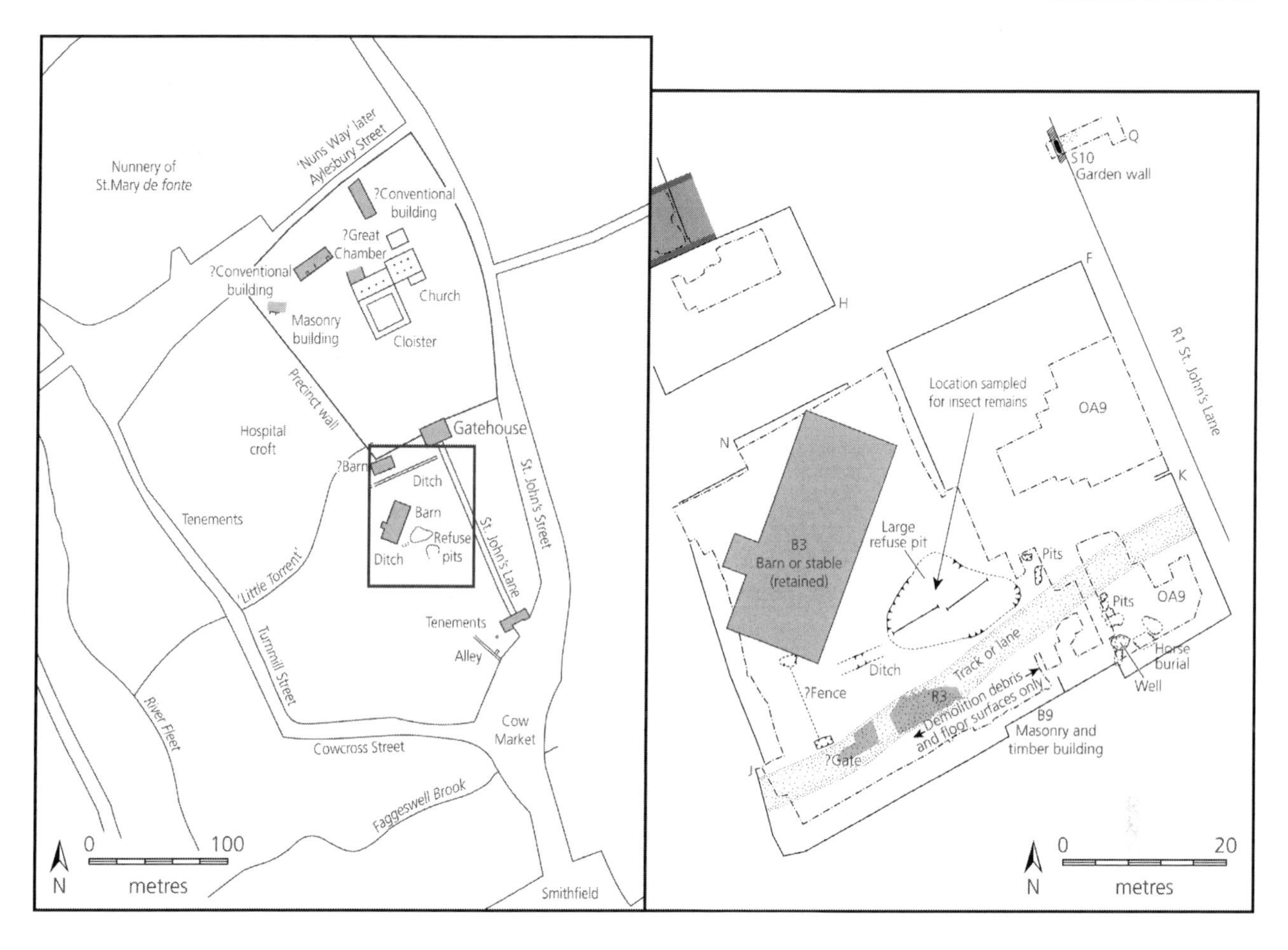

Figure 9.4. The precincts and the area of the barn and large pits from St John Jerusalem (redrawn from Sloane and Malcolm (2004)

developed to become a large precinct of religious houses with their associated lay structures (Figure 9.4) (Slone and Malcolm 2004). Most of the materials studied for insect remains came from the outer precinct to the south of the church. A large barn or stable, 12 by 25 metres, dominated this precinct during the 12th and 13th centuries (Sloane and Malcolm 2004). Around 5 metres to the southeast of this barn a large steep-sided pit, 18 by 12 metres had been dug. Originally this appears to have been a quarry pit used in the construction of the nearby buildings and was periodically full of water. Considerable quantities of material continued to be dumped into this feature until the late 15th century.

The biological materials from this pit are very striking in their nature and produce one of the clearest interpretations of all the deposits sampled in London. Anne Davis's (2004) study of the plant remains from this feature provided the first insight into the origins of this deposit. She found a range of seeds of typical of hay meadow plants such as buttercup (*Ranunculus* spp.), self-heal (*Prunella vulgaris* L.), plantain (*Plantago media* L.), field scabious (*Knautia arvensis* (L.) Coult.), daisy (*Bellis perennis* L.), dandelion (*Taraxacum* spp.), clover (*Trifolium* spp.), vetches (*Vicia spp.*), scented mayweed (*Matricaria recutita* L.), hawksbit (*Leontodon* spp.) and sedges (*Carex* spp.). In addition the deposit was also packed with grass stem and leaf fragments. Anne Davis felt that this probably represented hay, used both as bedding and fodder. In addition, cereal bran, pea pods and a range of arable and garden weeds were also recovered, suggesting that these were either fed to stabled animals or that other waste material had been dumped into the pit (Davis 2004).

The insect remains rather spectacularly, and very fortunately, confirmed this interpretation (Smith and Chandler 2004). Many of the weevils present are associated with the same range of hay meadow plants discussed above; clearly indicating that hay was a notable constituent of this deposit. Several species recovered, such as the *Typhaea stercorea*, the lathridiids and the cryptophagids are common in stored hay and bedding in stables (Smith 1998). Amongst the beetles a range of taxa, such the *Cercyon* species, some of the staphylinids and most noticeably the *Anthicus* species are associated with very rotten and 'hot' decaying material; commonly occurring in piles of stabling waste. A similar impression is gained from the flies, with *Sphaerocera curvipes, Ischiolepta* cf. *pussila* and cf. *Telomerina flavipes* present in large numbers in some samples. These flies are common in decaying stabling material (K.G.V. Smith 1989; Skidmore 1999). Equally, the presence of 'granary weevil' and the *Bruchus* 'pea weevils' confirms that spoilt grain and peas and beans were fed to the stabled animals. However, perhaps the clearest clue to the past use of this material comes from the lice recovered, in particular the heads of *Damalina caprae* the 'goat louse'. Given the known proximity of a nearby barn (Sloane and Malcolm 2004) I think we have a clear idea of what this material was and where it came from. Kenward *et al.* (2004) have recently published results from assemblages of insect and plant remains from a similar situation at Low Fishergate, Doncaster producing much the same interpretation.

WINCHESTER PALACE, SOUTHWARK

This large religious house on the south side of the River Thames was the residence of the Bishops of Winchester from the 12th century onwards (Seeley *et al.* 2006). The insect faunas examined come from a number of deposits associated with the inner courtyard and privy garden of the palace (Smith 2006).

Two 12th century pits contained notable insect faunas. One pit produced a range of species including the leaf beetle *Plateumaris sericea* and weevil *Thyrogenes*, both of which are associated with waterside plants such as sedges (*Carex* spp.) and rushes (*Juncus* spp.). This, along with water beetles, suggests that the pit was either flooded, or that rushes collected for flooring had been dumped at this location. A second pit of the same date produced another distinctive beetle fauna. It consists of a range of *Cercyon* species, the staphylinids *Lithocharis ochraceus, Leptacinus intermedius* and various *Monotoma* species. These species are thought to be typical of well-rotted piles of organic material, usually decaying stabling waste (Kenward and Hall 1995, 1997). This interpretation is supported further by the presence of pupae of the stable fly and the housefly. These are both common inhabitants of stabling waste (K.G.V. Smith 1989).

Perhaps the most significant set of insect faunas at Winchester Palace come from the stone lined pit at the junction of the east and south ranges. These deposits date to the 15th and 16th century (Seeley *et al.* 2006, 96). The fill of the pit consists of layers of dark brown organic silts and lenses of ash charcoal and cinders interleaved with white lime. The latter was probably used to seal and sanitise the pit. In terms of archaeology the use of the feature is clearly indicated by the presence of glass urinals (Seeley *et al.* 2006, 96). The plant remains are also very indicative. They contain the seeds of fig (*Ficus carica* L.), grape and raspberry or blackberry (*Rubus* spp.). These are seeds that tend to pass through the human gut undigested (you can research this phenomena yourself, particularly if you have small children) and enter deposits as part of cess (Giorgi 2006). A similar reasoning explains the presence of cereal bran in several of the deposits. However, the inclusion of food waste is also suggested by the numbers of fish bones, fruit stones, cereals and herbs such as coriander (*Coriandrum sativum* L.), dill (*Anethum graveolens* L.) and black mustard also recovered (Giorgi 2006; Rielly 2006). One notable inclusion is the almost complete skeletons of two cats. Rielly (2006, 154) explains their presence thus:

> *[this] either says something about deposition practices or points to the inability of London cats to avoid these traps to the unwary.*

This stone lined pit is of course a garderobe. The insect remains recovered support this conclusion and these will be discussed in detail when we go down the cesspit in Chapter 11.

PREACHER'S COURT, THE LONDON CHARTERHOUSE

The London Charterhouse, or rather the 'House of the Salutation of the Mother of God', used to lie just to the south-east of the site of the Priory of the Order of the Hospital of St. John Jerusalem and was in modern Smithfield just to the north of the city wall (Barber and Thomas 2002). Sir Walter Manny and Bishop Northburgh founded the religious house in 1371 perhaps as a response to the black death of 1349 since it incorporated at least two of the 'emergency' burial grounds associated with the plague (Knowles 1969; Barber and Thomas 2002). The house was Carthusian and intended to remain small in scale and to be contemplative in nature. It is suggested that it consisted mainly of a celled cloister, a small chapel and a number of cramped community rooms with a separate range for the lay brothers near the entrance to the precinct (Knowles 1969).

The insect faunas from this religious house date from three periods (Smith 2002). Sample 302 came from a pit that predates the foundation of the monastery. The remainder all come from a range of pits that were associated with a precinct wall that bounded the northwestern side of an inner court to the southwest of the great cloister (Barber and Thomas 2002). The pits and the deposits in them date from between 1371 to 1537. As was common with most of these monastic houses, 1537 represents the point when Henry VIII dissolved it, though in the case of the Charterhouse the final suppression seems to have been particularly harsh with the execution or starvation of the remaining monks and lay brothers (Barber and Thomas 2002).

The insect faunas recovered from these pits clearly suggest that they functioned as either cesspits or a location for the disposal of waste. The presence of food waste and cess is clearly suggested by the large quantity of fruit remains and bran frass recovered amongst the plant macrofossils (Giorgi 2002). The presence of food waste also is suggested by the abundant remains of animal bone, fish bone and oyster shell (Barber and Thomas 2002). Settlement rubbish, if not craft waste, also seems to have been dumped into the pits based on the large amounts of leather off-cuts encountered in one of the pits. The insect remains recovered suggest that cess was present in some quantity (Smith 2002). Again the nature of these insect faunas will be considered further when we discuss cesspits in Chapter 11.

CONCLUSION

From the above we can see that the Medieval insect faunas of London are from that type of deposit that has previously been described as a rarity, an unmixed deposit placed *in situ* and which provides a clear interpretation. This leads us nicely on to the issue of how we interpret such deposits and the various intellectual pitfalls that exist.

But before we discuss this, there is one further issue to consider. Does this quick survey of the insect life of a range of Medieval monastic buildings and communities suggest that the disposal of stable waste, domestic waste and cess was a pre-eminent occupation? This is, of course, complete nonsense. There is more to life than cesspits. If we return to some of the considerations we raised in Chapter 1 over the formation of the archaeological record, in the case of Medieval London the over-representation of waste and cesspits probably results from the way that such deposits get preserved and enter the archaeological record rather than anything else. It is a virtuous circle. Insect remains require waterlogging and the most frequently encountered waterlogged deposits on urban Medieval sites are cesspits, because of the depth of these features. As a result, the recovery of waterlogged deposits from features other than cesspits in Medieval London (and elsewhere for that matter) should be a research priority.

CHAPTER 10: DEFINING INTERPRETATION GROUPS: ARCHAEOENTOMOLOGY COMES OF AGE

INTRODUCTION

This chapter takes a slightly different approach from the others in this book. It concentrates on what I consider to be the main development in the approach to archaeoentomological analysis over the last decade (mid 1990s – mid 2000s). This is the development of a number of clear groups of insects that can be used for direct archaeological interpretation. This has completely advanced how we think about interpreting insect faunas from archaeological settlement because:

1) it helps us to resolve some of the issues, especially contradictions, raised in the late 1970s and early 1980s. These had become a major difficulty for the discipline. Not only were we left with a lack of confidence in our interpretations, but this often resulted in a 'muddled' message for research collaborators and commercial clients.
2) the groups outlined are derived directly from the archaeological record rather than relying on modern ecology or analogues. This functions to reduce the number of 'interpretative hoops' we have to jump through and deals effectively with a lack of applicable modern analogues for the deposits we study.
3) there is now deliberate integration of archaeoentomological data with other environmental indicators such as plant macrofossil, animal bone and parasite remains as well as 'standard archaeology' itself. In the past the combined use of differing disciplines (or a multi-proxy approach) to look at the interpretation of a single deposit has either been completely absent or implicit rather than explicitly detailed.

A WAY FORWARD: INTERPRETING URBAN INSECT ASSEMBLAGES

In the introduction to this book, I outlined a simplified history of the development of urban archaeoentomology. I said that I would return to this discussion later in the book and, I am afraid, this is the time and the place.

Kenward in his 1975 *'Pitfalls in the environmental interpretation of insect death assemblages'* paper and his 1978 book *'The Analysis of Archaeological Insect Assemblages: A New Approach'* raised a number of issues that openly reviewed flaws in archaeoentomological interpretation of insect remains from urban deposits. Firstly, we are not dealing with a living insect fauna, or community, but rather a death assemblage formed as a result of how the material was used, deposited and decayed. The key problem identified by Kenward (1975b, 1978) was the unpredictable inclusion of 'allochthonus' fauna, (species that enter deposits in flight, on foot or by accident). These species are not strictly indicative of the deposited material or conditions 'on site' (Kenward 1978). Another issue is the fact that populations of some species can be 'superabundant' in deposits and occur in much larger numbers than others. This might mean that they obtain an importance they do not really warrant and may actually distort the interpretation. A clear case in this study is the faunas that are dominated by grain pests. If we relied on the insects alone, we might see these deposits as pure grain, but often other lines of evidence may indicate that the deposits contained much more mixed materials than this. Furthermore, many modern ecological records may be inappropriate for us to understand the behaviour of these species in the past. We do not have modern analogues for things like damp earthen house floors and buried food waste. Kenward (1978), therefore, suggested that we needed to develop a new set of interpretative tools if we are to consider these faunas in a meaningful way.

Know the type of insect faunas you are working on

In particular, archaeoentomologists (but really all archaeologists, environmental or otherwise) need to consider their deposit's origin and whether is it safe to interpret it. This is, basically, a very simple idea. An archaeological insect fauna with many individuals of a limited number of species probably represents a deposit that is *in situ* or results from one human action. Similarly, if the majority of the faunas are from one type of ecology again there is probably a single origin for the material. These types of fauna are, of course, reasonably safe to interpret. I suppose the clearest examples from this book are the various deposits of Roman grain waste discussed in Chapter 5 and the example of the 'big pit' from The Priory of St. Johns of Jerusalem in the previous chapter.

There also are faunas where a large number of species are represented by small numbers of individuals. Often this type of fauna contains species from widely different ecosystems. Faunas such as this probably represent very 'mixed' deposits, those that have been re-deposited or that have collected large numbers of species from the surrounding environment. Classic examples are the faunas from the Guildhall outlined in Chapter 5 and a range of faunas from wells where there has been no deliberate dumping of settlement material (Hall *et al.* 1980; Simpson 2001).

Kenward (1978) suggested that a range of statistics could help outline whether deposits are safe to interpret. These include the use of rank order curves and diversity statistics – particularly Fisher's alpha (α), and the percentage of the outdoor fauna present. Over the years I have found that many of my students get unnecessarily panic-stricken by these 'stats' and fail to see them as a means to an end. To be fair, I have only quite recently realised that they are often no more than a 'middle range' reality check before you start to place your faith on any interpretation of a fauna. Personally, once the dangers and principles of the 'pitfall' problem (Kenward 1975b) were understood I have tended to do this 'by eye' rather than run the statistics. Of course the role that assessment procedures and 'semi-quantitative scans' of insect faunas have in helping with this problem have been discussed in Chapter 8 and elsewhere (Kenward *et al.* 1985; Kenward 1992).

Use analogue studies designed to deal with archaeological rather than ecological questions

This involves collecting modern samples that replicate actual archaeological situations and analysing them in order to better understand the nature of the insects present in 'archaeologically comparable' materials. It has been shown to be very useful as a way of understanding how archaeological insect faunas are formed. It was this approach that raised the whole issue of the introduction of allocthonous faunas in the first place (Kenward 1975b, 1978). It also can produce ecological information that is directly relevant to archaeoentomology.

During the 1990s I spent a lot of time investigating modern analogues for the archaeological record. In particular I investigated hay and stabling materials (Smith 1998, 2000a) and roofing thatch (Smith 1996a, 1996b; Smith, Letts and Cox 1999; Smith, Letts and Jones 2005). In the main I found that these studies tended to generate more questions and identify even more problems in terms of the way that insect remains may enter the archaeological record than they actually solved. At times I felt like an 'Elijah', since I usually said what we could not do rather than what we could. I will return to address how these studies have coloured my understanding of archaeological insect assemblages in the discussion below, when relevant.

Use other orders of insects

Kenward (1978) urges the palaeoentomologist to think beyond the beetles and to consider the role that other insect orders can play in archaeological interpretation. I have already discussed a number of occasions when lice and fleas can play a decisive role in the interpretation of a deposit. One final example from the previous chapter is the goat lice from the 'great pit' at the Priory of St. John of Jerusalem.

Fly pupae have also been found to have a crucial role (e.g. Panagiotakopulu 2004). Certainly, for many of the deposits studied in this book their interpretation did not become clear until I also started to regularly identify the fly remains from the samples (since 1997). If you are after a future career in archaeoentomology (as well as forensic science) where you would always be in demand please consider specialising in the flies. I can just about handle the common pupae, but the less orthodox and the adult remains are still beyond me.

Develop 'interpretative' groups of insects based on the archaeological record not modern ecology alone

In his 1978 volume on the interpretation of archaeological insect faunas Kenward suggests that interpretation needs to be based on the 'whole fauna' rather than its individual elements. One way forward, he suggested, would be to develop descriptive groupings based on a most general understanding of their modern ecology. This would allow simple comparisons to be routinely made between deposits. Initially these consisted of only an 'outdoor grouping' (group OD) and an aquatic grouping (group a or W) and were intended to help understand the proportion of allochthonus fauna in a death assemblage (Kenward 1978). By 1982 the three compost groupings (group rd – rotting dry, group rf – rotting foul and group rt – rotting general) had been proposed (Kenward 1982). These groupings were used, along with a category for species from dead wood (group L) and grain (Group G) by the time of the publication of the Lloyd's bank site and the report on the preliminary sections at Coppergate (Hall *et al.* 1983). They subsequently became formalised in the reports on the Colonia and Coppergate assemblages (Hall and Kenward 1990; Kenward and Hall 1995). These groupings have now been widely used and are an intrinsic part of most archaeoentomological site reports. Indeed, they have been used at times in this book. The membership of each grouping, is generally outlined in Figure 1.2. However, we must be clear on this point, these groupings were only ever intended to be a descriptive guide and somewhat of a stopgap. It was also clear, even at the earliest stage of rumination, that these groupings could be re-defined so that they are more directly applicable to the archaeological record itself.

At this point we begin to see a move towards using the archaeological record directly to define groupings of insects that are useful for archaeological interpretation. At first, this may seem to be a prime example of circular logic but its application has major advantages providing you can demonstrate that the groups are coherent, reliable and applicable. I think we need to be clear about what has happened here. We are now no longer relying on modern ecology to interpret insect faunas. We are now primarily using the archaeological record, with modern ecology serving as a check. This is a major shift in how we interpret this material or indeed think about insect faunas from archaeological sites.

In order to start this process Kenward (1982; Hall *et al.* 1983) undertook a series of comparative statistics on the assemblages from Medieval Coppergate and Lloyds Bank, as well as Roman Skeldergate sites, in York and Medieval Beverly. In this case a simple statistic called Jaccard's coefficient was used to see how often pairs of species occurred together and how strong any connection was. Kenward displayed the results as 'constellation diagrams' where pairs of species that always occur together are arranged nearby and the strength of the relationship indicated by the number of lines connecting them. The results from this initial work are shown in Figure 10.1.

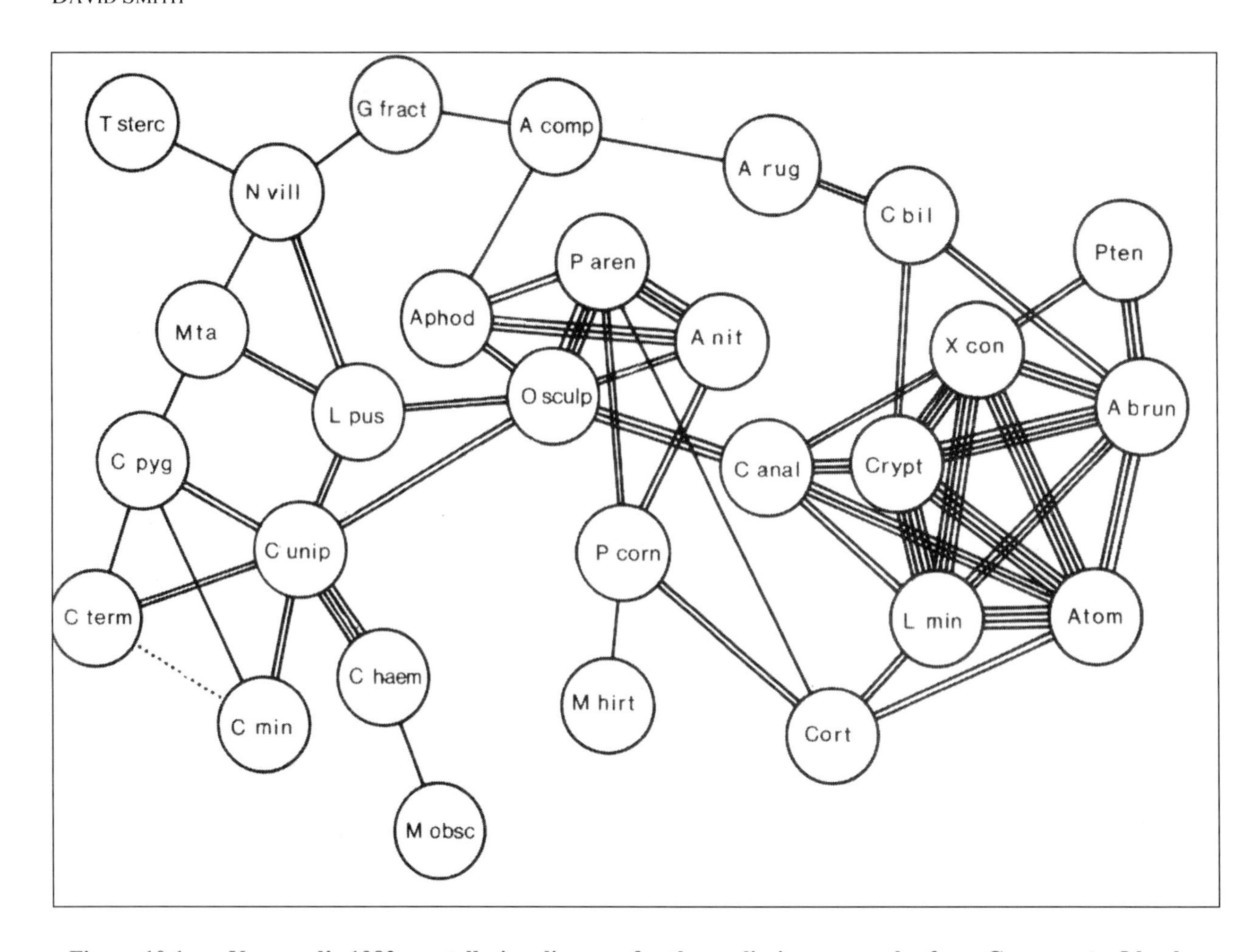

Figure 10.1. Kenward's 1982 constellation diagram for the preliminary samples from Coppergate, Lloyds Bank, Skeldergate and Beverly (reprinted with permission from Kenward 1982).

This suggested the existence of a number of 'communities' or groupings. What must be made clear is that this is not necessarily the arrangement of species we might have predicted based on their modern ecological behaviour. It was at this point in time that some of the strange juxtaposition of species from apparently contradictory ecologies discussed in Chapter 5 became apparent. This initial statistical analysis lead to the suggestion (Hall and Kenward 1990; Kenward and Allison 1994; Kenward and Hall 1995) that three assemblages (or pseudo-communities) of beetles commonly occurred in materials from urban archaeological sites, in addition to the groups based on modern ecology discussed above. These were the 'house fauna', the 'oxyteline group' and the 'subterranean/ post-deposition community' (Kenward and Hall 1995, 464). The common members of these groups are listed in Figure 1.2. Where this exercise became exciting was that Kenward was able to return to the archaeological record at Coppergate with a fresh eye. He found that the 'house fauna' was commonly associated with the interior of buildings; the oxyteline group frequently is associated with open areas, such as yards and pathways where wet mud probably dominated and the 'post-deposition community' is associated with sealed fills in pits and dumps.

Subsequently, Carrott and Kenward (2001; Kenward and Carrott 2006) have run a comparative exercise on 750 samples from the Coppergate site. The process was similar to that that already described but used Spearman's rank order coefficient to identify recurrent pairs of insects and the strength of the relationship between them. The 'constellation diagrams' are displayed in Figure 10.2 and the main members of each group and their possible interpretations are discussed in Figure 1.2. The strength of these groups was also independently 'tested' by carrying out a detrended canonical correspondence analysis (DCCA) on a selected group of taxa from these sites (the use of DCCA) will be discussed below). When the membership of the proposed groups were plotted onto the resulting data cloud of individual species it was clear that the species from the various groups tended to plot out together (this can be seen in Figure 10.3.).

Carrott and Kenward (2001) then returned to the archaeological record for the Coppergate site. They found that many of the groups identified tended to relate to specific types of deposits such as internal settlement deposits and floors, cesspits, buried rubbish, gullies and wet external areas, and very foul deposits in pits.

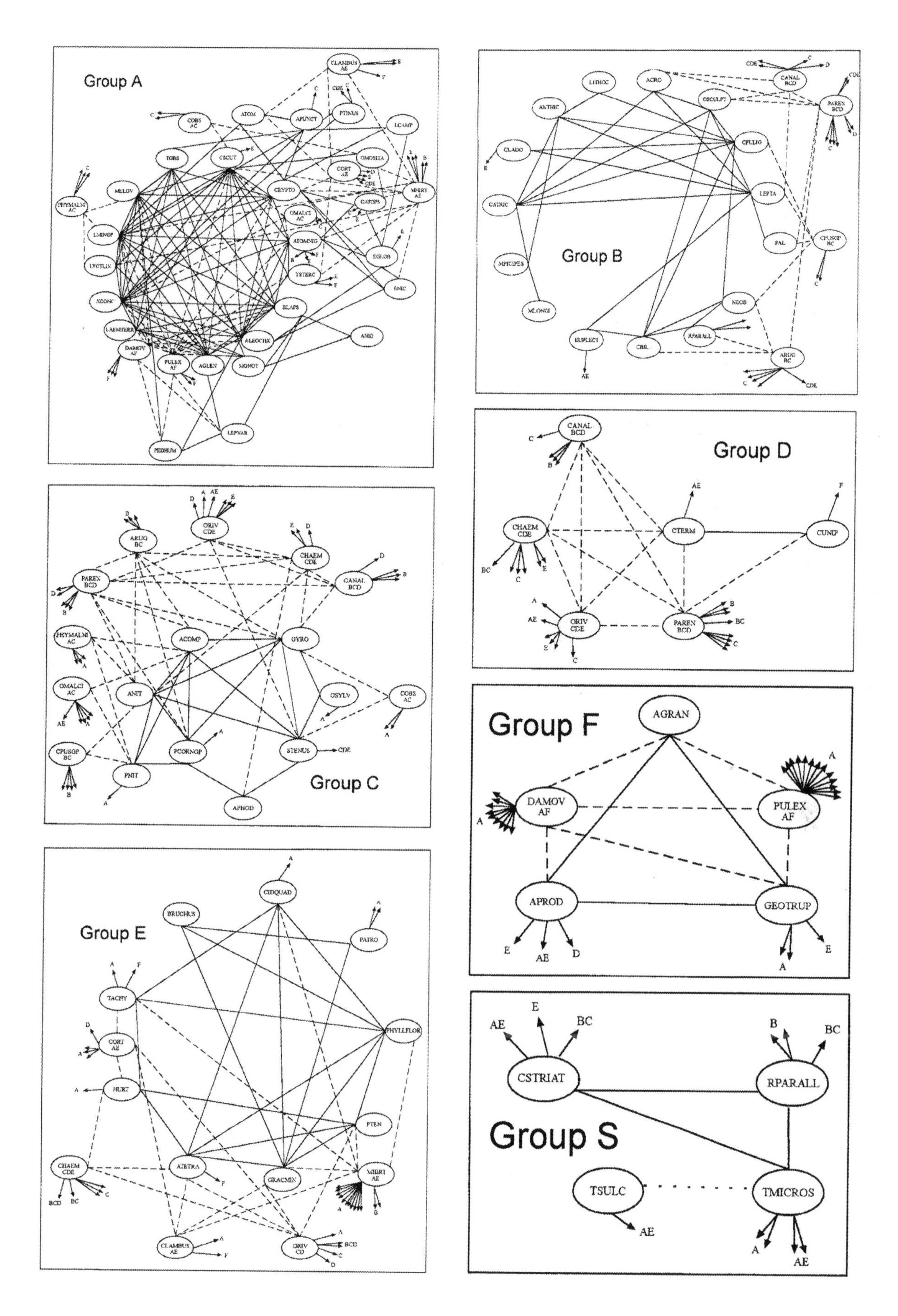

Figure 10.2. Associations of species from Anglo-Scandinavian deposits at Coppergate York based on the Spearman's rank order correlation (reprinted with permission from Carrot and Kenward 2001).

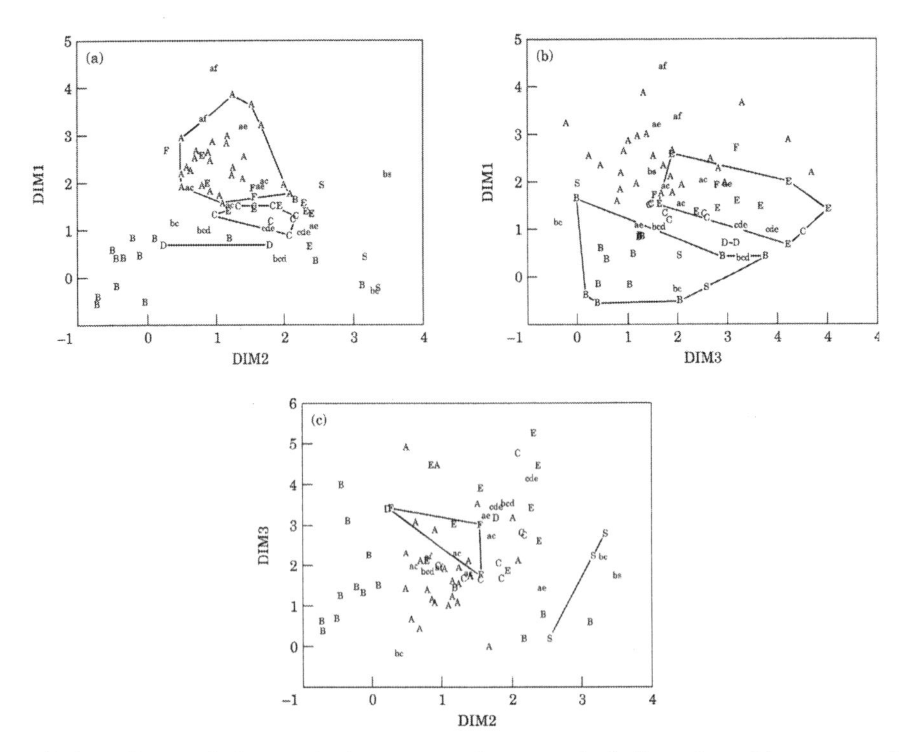

Figure 10.3. Detrended canonical correspondence analysis for selected insect taxa in archaeological assemblages from Anglo-Scandinavian deposits at Coppergate, York. Comparisons of the 3 axes of variation are plotted and the taxa have been assigned to habitat groups. (Reprinted with permission

Recently, Kenward and Carrott (2006) have undertaken a similar set of analyses on a wider range of sites from Northern England and Ireland. They found that the groups were more-or-less applicable across the range of sites considered. However, the membership of groups did change subtly between the sites. According to Kenward and Carrott (2006), this reflected the character of the sites. In the main the difference seems to be between sites where contexts and assemblages are clear and ones in which they are not. We will return to the issue of how 'mixed' sites are formed as the archaeological record develops and the extent to which sites are 'turned over' by human action, but first it might be worthwhile to consider some of the dangers of this approach.

THE DANGER OF JUST USING INSECTS GROUPS

With Kenward and Carrott's work we possibly have a set of 'indicator groups' that can be applied widely across archaeological sites with their interpretative strength coming from repeated recovery on site and the clear links to specific types of archaeological contexts and situations. However, there remains great danger in the unquestioned use of such groups. Kenward's 'new groups' are of course based on the situation at York. It could be dangerous to just transfer them wholesale to another site, where different conditions and depositional problems came into play. The dangers can be seen in a number of analogue studies I carried out as part of my PhD dissertation research at Conisborough Parks Farm, South Yorkshire.

The 'big idea' for my PhD was to look at how the insect fauna of hay and straw changed as this material moved from the field to store, then into the cattle pen and finally onto the farmyard midden (Smith 1998, 2000a). This also turned out to be a 'bad idea' at the same time. It meant that I had to dig my way through metres of cow-fouled straw at least once a month for a year. The real high spot was when one of the cows unexpectedly licked an extremely delicate part of my anatomy whilst I was bent over a hole I was digging into the byre. The unexpected bovine caress of a relatively delicate area meant I ended up propelling myself head first into a half metre of stale cow urine. I remain deeply psychologically scarred to this day.

Anyway, the important point is that at the start of the study we felt that each of these stages in the movement of material around the farm would produce a distinct set of insect faunas. It quickly became apparent that the story was not so straightforward. This seemed to be for three reasons:

1) many live and dead individuals were carried from one 'stage' to another. This complicated the pattern produced, resulting in situations similar to the

insect faunas from 'mixed' sites as discussed in Chapter 8.

2) some stages were not represented by any fauna at all, notably in this study, the cattle pens themselves. In this case the lack of insect life probably results from the presence of so much stale ammonic urine. It was almost the end of me as well.

3) the insect faunas present are not indicative of the actual material but rather the general 'environmental conditions' present (e.g. humidity, relative warmth, etc…).

The last point needs to be expanded on. The species we might standardly use to indicate the presence of dry hay (for example the 'barn beetles' *Typhaea stercorea*, *Enicmus minutus* (group) and cryptophagids) are not actually representative of the 'hay' *per se*. They actually only indicate the presence of fairly dry plant material with a certain level of humidity and mould growth. The problem is that these conditions also occur in several different types of material scattered around traditional human habitation.

A classic example of this can be seen in another part of the study I undertook at Conisbrough Parks Farms. In this case, samples were collected to examine the nature of the farmyard midden. At one stage in the summer this pile dried out to such an extent, mainly as the result of the internal heat generated from biological breakdown, that the conditions again became favourable for the 'barn beetles'. Does this matter? Yes, since misinterpretation can easily result.

Perry, Buckland and Snæsdóttir (1985) undertook a cluster analysis of the insect faunas from a Late Medieval farmyard midden from Stóraborg, Iceland. The analysis identified three clear groups of faunas. There was a 'hay residue' that consisted of the 'barn beetles', a 'house floor fauna' which consisted of a range of species associated with foul matter dominated by *Catops fuliginosus* and a blend of the two types of fauna. Perry, Buckland and Snæsdóttir interpreted this as indicating that the material in the midden came from different locations around the farm, notably, the hay store and a range of dirty house floors. This is, of course, entirely possible. The problem is that these faunas could also be explained in another way. Recently, Buckland (pers. com.) has suggested that 'the dirty house floor' fauna may actually be indicative of Icelandic hay meadows (hence the presence of *Catops fuliginosus*) and used stabling matter. If this is the case then it is possible that the presence of the 'hay residue fauna' (a.k.a. 'barn beetles') in these samples from Stóraborg could result from the development of similar conditions in the hot interior of the farm midden as were seen at Conisbrough.

The difficulties result from the fact that insect faunas often represent conditions, rather than specific materials. This also can be seen when we consider thatch. A series of studies undertaken by a number of students and myself on thatch roofs clearly suggested that a distinct fauna is repeatedly recovered from this type of material (Smith, Letts and Cox 1999; Smith, Letts and Jones 2005). However, it could not be demonstrated that the environmental conditions that favoured this community are limited to roofing thatch and could not occur in other settlement materials. Of course the answer to this problem is easy. Don't *just* rely on the insects.

INDICATOR PACKAGES

In 1997, Kenward and Hall (1997) formalised an idea that had been in general use but not stated. That the interpretation of the biological remains from archaeological sites is best achieved by using 'suites' of evidence derived from all available sources of information. It does not matter if these interpretations are based on insects, plant macrofossils, pollen, parasite ova, animal bone or other elements of the archaeological record itself; the important thing is that all the evidence points towards the suggested reconstruction. Where the evidence appears to combine commonly in the archaeological record, and where the interpretation reached appears to be consistent, then an 'indicator package' could be put forward. If this package re-occurs elsewhere in the archaeological record, assigning a similar interpretation should be straightforward. Moreover, if the insect 'indictor group' for this 'package' came up I could then ask the other specialists involved in a project if they had the other components of the 'package'. The first example that Kenward and Hall chose to use was that of stabling material (Kenward and Hall 1997). Subsequently, they have suggested that similar packages exist for a range of craft and production activities (Hall and Kenward 2003).

Lisa Moffett and I also tried to play the same game with roofing thatch (Moffett and Smith 1997). Earlier I said that the difficulty with using the apparently distinctive fauna of insects seen in roofing thatch to identify the presence of this material in the archaeological record was that it could occur elsewhere. I also suggested that if it were to be used in such a way, other forms of independent evidence would be needed. When I looked at the material from the Late Medieval tenement in Stone, Staffordshire, I found that the insect fauna was essentially the same as in the modern thatch roofs. I asked Lisa, who was analysing the plant remains from this site, if she had any evidence that pointed in the same direction. It turned out that rye (*Secale cereale* L.) straw, along with other unidentified cereal straw, dominated the plant remains recovered from this sample. Rye straw (indeed many types of cereal straw) was a common type of thatching material in this period (Letts 1999). Lisa then examined an unprocessed sample of material from this deposit and found that the cereal stems were closely packed together tended to lie in the same direction and closely resembled thatch. Similarly, the archaeologist felt that the stratigraphy could indicate a collapsed roof. All in all it led us to 'tentatively' suggest the presence of a thatch roof (though heavens knows why we were so tentative, if it looks like a duck, quacks like a duck, etc….). I suppose that the final proof would have been the presence of spars or ties used in thatching from the deposit, unfortunately, these were not recovered at Stone.

In Chapter 4, when discussing insect faunas from the wider landscape, I raised a number of issues that prevented us using insect remains to 'read' the nature of the landscape in a straightforward 'mosaic' way. I suggested that the end result, to borrow Caseldine's terminology, was that we could not expect to produce reconstructions that were 'truthful' but rather we work to produce 'plausible' reconstructions. The key was to judge which possible reconstruction was more plausible. I think we can see the development of the interpretation of insect remains, through its differing stages (i.e. reliance on individual species as part of a 'mosaic', questioning extent of allochothony, using ecological groups derived from modern ecology, using groups derived from the archaeological record, and the inclusion of such groups into 'indicator packages') as a progressive move towards making the interpretation of insect faunas from urban sites as reliable as it can be.

CHAPTER 11. A DETRENDED CANONICAL CORRESPONDENCE ANALYSIS (DCCA) OF THE DATA FROM LONDON AND COMPARISON WITH COPPERGATE, YORK

INTRODUCTION

In an attempt to see if any patterns in the London data were discernable between sites and periods it was decided to carry out an ordination of the data available. In addition, it would be nice to see if the recent interpretive groupings of Carrott and Kenward (2001; Kenward and Carrot 2006) are as applicable to the archaeology of London as they were to that of York. The ordination chosen was detrended canonical correspondence analysis (DCCA) using the CANOCO 4.5 computer package (ter Braak and Šmilauer 2002).

Ordination of data is a commonly used technique that allows us to search data for patterns. Ecologists with an interest in biological communities often use it (e.g. Gauch 1982; ter Braak 1987) as well as archaeobotonists (e.g. Jones 1991; van der Veen 1992; Bogaard 2004) and archaeoentomologists (Carrott and Kenward 2001; Whitehouse 2004, Kenward and Carrott 2006). The statistical technique used to carry out the ordination is based on reciprocal averaging (ter Braak 1987) but for most of us what is of interest is that it produces ordination diagrams that place samples with statistically similar contents (or indeed species with statistically similar behaviour) closer together in the plots and samples with different contents further apart. This enables a quick and independent check of patterns one can intuitively 'detect by eye' in the data. The ordination can then be plotted and the various groups of samples or species can be labelled using a set of independent categories (for example you could use Kenward's ecological codes).

DATA SELECTION AND MANIPULATION

I initially chose to use the whole of the available dataset from London. That is, all 394 individual taxa, all 131 samples and all 17,476 individual insects. This produced a blizzard of data and resulted in both the plot of the ordination of the species and the plot of the samples showing no discernable pattern. This is a common problem. Reciprocal averaging will tend to give rare species, or samples with a low abundance of individuals, undue importance and result in a confused diagram (Gauch 1982).

I therefore decided on drastic action. I excluded any sample with less than 50 individuals from the analysis. This sounds drastic but in the normal day to day business of archaeoentomology I would tend to regard samples with this low a number of individuals as potentially unrepresentative. Indeed, if I were assessing the samples commercially I would probably recommend that no further work occurred on faunas with such a low number of individuals.

I also excluded species from the analysis who were present in less than 10% of the samples. In effect, any species that did not occur more than 13 samples out of the entire 131 sample dataset was removed.

CANOCO enables you to use a number of techniques to transform the data. This includes using the square root of the numbers and a logarithmic transformation. This aims to smooth out any natural statistical difference between samples where a large number of taxa occur in small numbers when compared to samples were a single species is superabundant. Though both types of transformation were tried during this analysis they proved to be completely unnecessary and in fact 'flattened' the plots. The only option that was selected, and ultimately retained, was that rare species were 'down weighted' and so were given less statistical importance.

THE LONDON DATA ORDINATED BY SAMPLE TYPE

The first question to explore is whether the type of context sampled influences the data pattern. In other words, are the insects recovered from house floors distinct from those recovered from wells? Figure 11.1 shows the ordination of London assemblages in terms of the type of archaeological context sampled. There is some loose clustering of specific context types; however, in nearly all cases many samples are plotted well away from any 'main group' for a particular context type, there is also a high degree of overlap between the types of context.

In general, external yard and dump deposits from sites such as Saxon and Norman Guildhall appear to plot out on the left of the data cloud. These also tend to be contexts where we know the insect faunas appear to have been 'mixed' as they formed. Most of the contexts thought to be from cesspits, and a number of other pit fills, tend to plot out to the right. Internal floor deposits appear to cluster toward the lower centre of the data cloud. A number of samples from pit fills, dump and road dump deposits are plotted in the upper part of the data cloud. These tend to be faunas that are dominated by the remains of grain pests, and mainly come from Roman Poultry and the late Medieval contexts from St. Mary's Clerkenwell.

The vast majority of the contexts, regardless of their categorisation by the archaeologists, fall into a jumble in the middle of the data cloud. These all tend to be the internal floor deposits from Bull Wharf, a lot of the pit dump deposits from Poultry and the contents of the 'big pit' at St. John of Jerusalem. I think that this just goes to prove an established point; the vast majority of urban archaeoentomological faunas are variations on a common theme; centred on housing, stabling and settlement.

THE LONDON DATA ORDINATED BY PERIOD

The next logical question to explore is whether the period of a deposit produces a distinctive assemblage. Obviously, certain periods produce an abundance of distinct sample types (e.g. cesspits are unique to the Late Saxon and Medieval periods) that could influence the data of a particular period; nevertheless, there might be

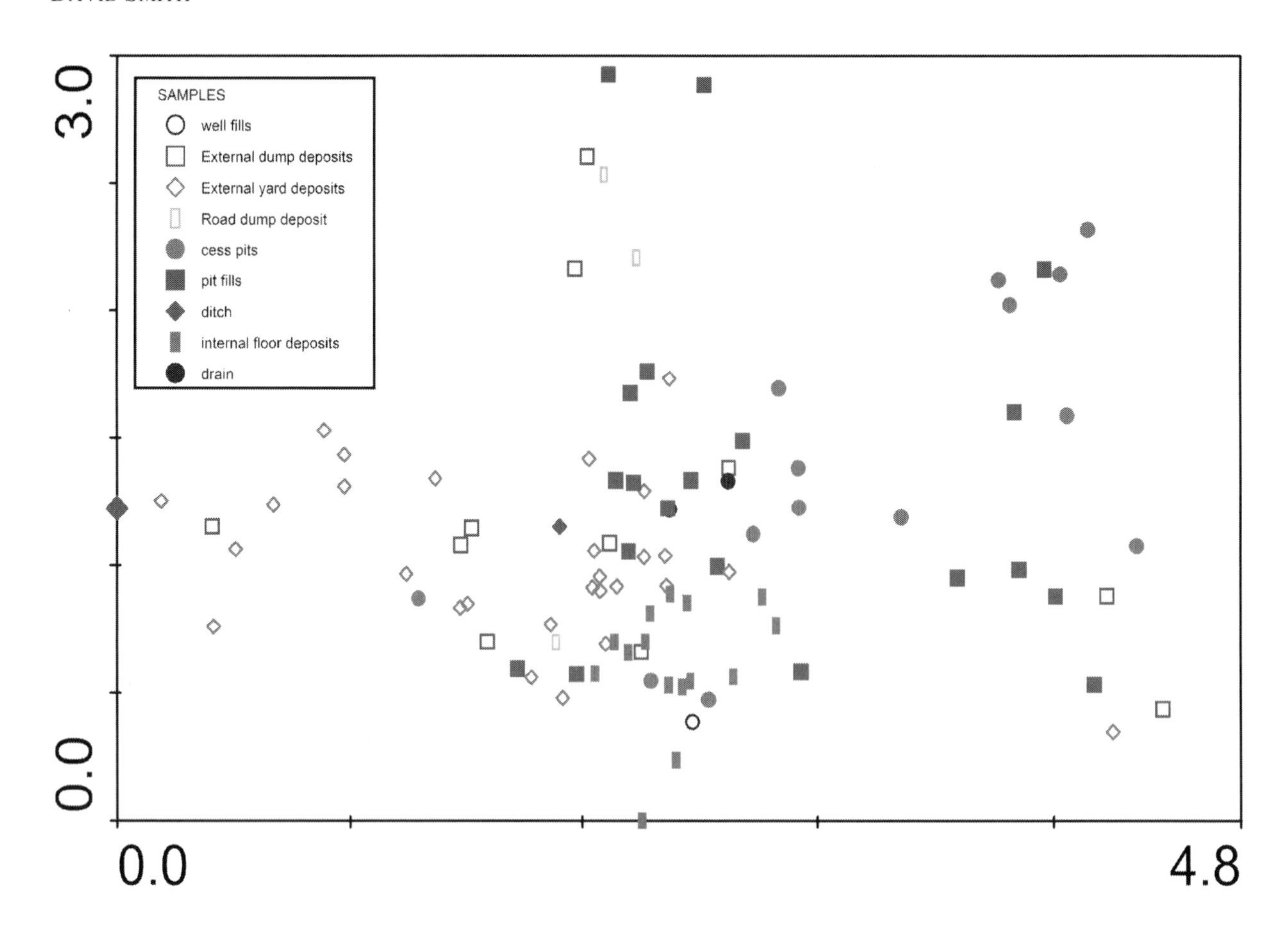

Figure 11.1. The results for the CANOCO DCCA ordination of London archaeoentomological data classified by sample type

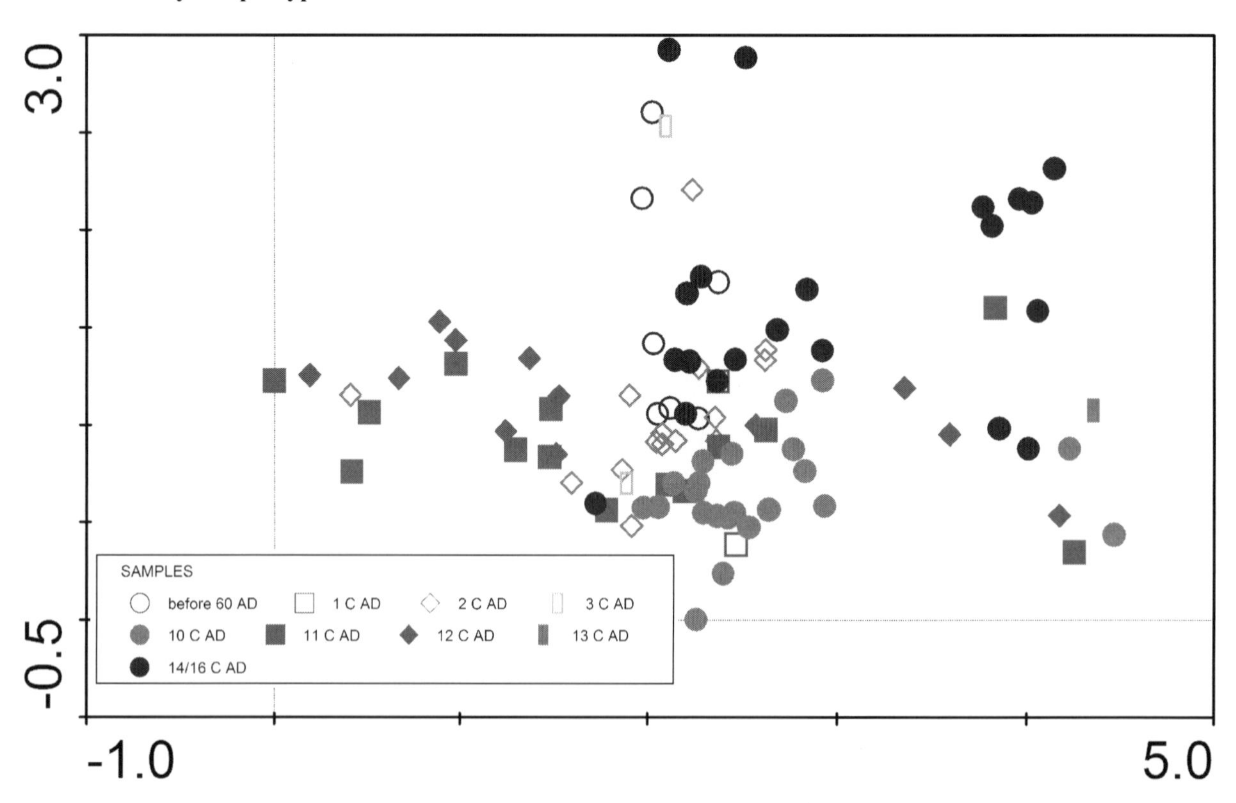

Figure 11.2. The results for the CANOCO DCCA ordination of the London data classified by sample date

distinctive patterns for Roman or Medieval insect faunas recovered from London. Figure 11.2 is the same plot as Figure 11.1, but in this case the various ages of the deposits have been indicated. There is no pattern here at all. Given the first result, this is not a surprise. Many urban faunas from vastly different time periods are essentially the same in London, as has also been noted for York (Kenward 1982; Kenward and Hall 1995; Kenward and Allison 1995; Kenward 1997). Provided we are working with urban deposits there tends to be no variation across time (grain pests being an exception).

THE LONDON DATA ORDINATED BY SPECIES

If sample type and period do not influence data, could individual taxa be influential? Figure 11.3 shows the London data ordinated by species and these are then labelled using Kenward's (Kenward and Hall 1995) groups, including the 'house fauna', 'subterranean' and 'oxyteline' groups. To my infinite relief, the DCCA ordination actually produces a relatively convincing set of groups:

1) **settlement/ dry**: in the middle of the data cloud is a group that consists of the majority of the 'house fauna' including *Xylodromus concinnus, Cryptophagids, Enicmus minutus, Anobium punctatum,* and *Aglenus brunneus* along with species such as the hide beetle *Omosita colon* and the 'subterranean' *Trechus micros* and *Rhizophagus* species. This clearly confirms the strength of Kenward's suggested 'house fauna' and increases the possibility that this is a real ecological community. It is interesting to see that *Oxytelus complanatus* falls out into this group, since I have always suspected that it is probably semi-synanthropic and Kenward holds it is common in the house floors at York (Kenward *pers. com.*)

2) **settlement/ dry**: in the middle of the data cloud is a group that consists of the majority of the 'house fauna' including *Xylodromus concinnus, Cryptophagids, Enicmus minutus, Anobium punctatum,* and *Aglenus brunneus* along with species such as the hide beetle *Omosita colon* and the 'subterranean' *Trechus micros* and *Rhizophagus* species. This clearly confirms the strength of Kenward's suggested 'house fauna' and increases the possibility that this is a real ecological community. It is interesting to see that *Oxytelus complanatus* falls out into this group, since I have always suspected that it is probably semi-synanthropic and Kenward holds it is common in the house floors at York (Kenward *pers. com.*)

3) **stabling/ wet ground:** in the lower part of the diagram there is a group of species that mainly consist of members of Kenward's 'oxyteline association' such as *Oxytelus nitidulus, Trogophloeus bilineatus, T. fuliginosus* and *T. corticinus*. In addition the various *Lithocharis* and *Leptacinus* staphylinids also occur in this area of the plot. Kenward has specifically linked all of these taxa to stabling deposits in the archaeological record (Kenward and Hall 1995; Kenward and Hall 1997). Again, their association here in the London material might suggest that they are a real community rather than an archaeological trend. Certainly the fact that the stable fly *Stomoxys calcitrans* falls into this same area of the data cloud supports this argument.

4) **foul/ rotting:** just above the 'stabling/ wet ground community is a rather more nebulous group of species that includes many of Kenward's 'rotting foul' group. This includes the *Cercyon*, and the *Aphodius* dung beetles along with *Oxyomus silvestris, Trox scaber* and *Platystethus arenarius*. Most of these species frequently are associated with rather loose wet and very decayed plant matter and wastes. *T. scaber* usually is associated with detritus in birds nests. In terms of the flies, both the housefly and *Sepsis* species fall into this area. This clearly confirms the rather horrid nature of the material probably favoured by this group of species. Again this analysis appears to validate Kenward's proposed 'rotting/ foul' community.

5) **grain pests:** at the top of the data cloud is a very clear group of species that are all grain pests. This is not a surprise but it does at least function as a check for how successful the ordination has been and strengthens the argument that the groupings seen are 'real' and represent plausible communities in the archaeological record.

6) **decaying material:** there is also a group that I have labelled 'decaying material'. In some ways this represents all that is 'left over'. However, it also contains several species of weevil, such as *Sitona, Gymnetron* and *Apion* that are commonly associated with field hay. Also in this area of the plot is *Typhaea stercorea* a species that is common in dry plant matter and hay. Other species roughly plot out in this area are the *Anthicus*, which are normally associated with the hot interiors of farmyard middens. Similarly the fly *Sphaerocera curvipes* is often associated with stabling matter (K.G.V. Smith 1989). I am not completely convinced, but we could perhaps see this grouping of species living together in something like hay-based waste and stable matter. It seems likely that the conditions within this material can vary greatly (e.g. along a spectrum from wet to dry, clean and fresh matter to foul and rotting, etc...) so it is possibly unsurprising that this group does not plot out neatly. Moreover, the multiple uses of plant

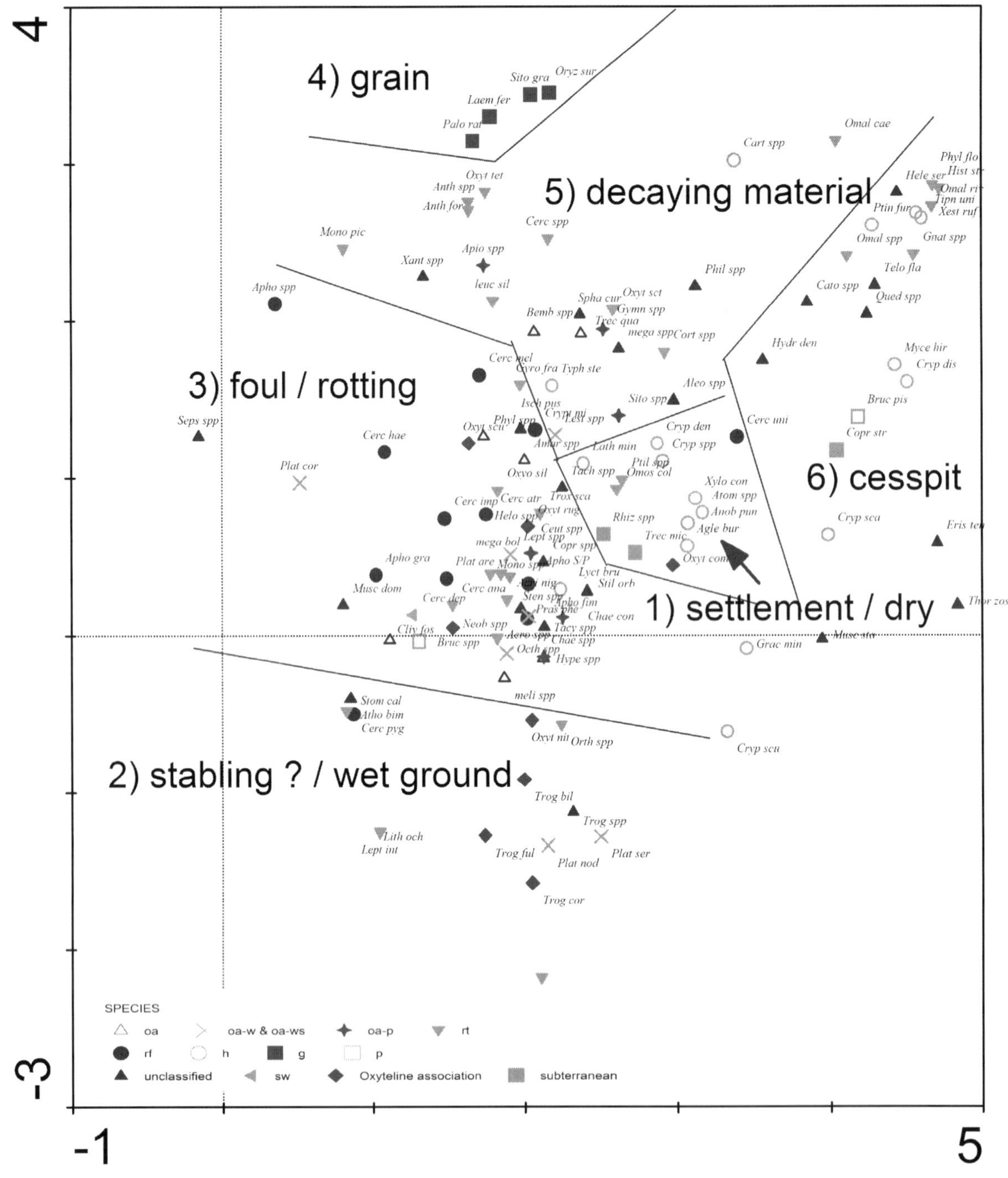

Figure 11.3. **The results for the CANOCO ordination using the samples from London using the ecological codes from Kenward and Hall 1995.**

materials such as hay and straw in the past may further affect archaeological results.

7) **cesspit:** finally, there is a group of species over on the right-hand side of the data cloud that clearly belong together. I have labelled this part of the plot 'cesspit'. However, since this seems to be the main 'big result' of this analysis of the London material I reserve the right to discuss it below.

Figure 11.4 shows the same ordination but in this case Carrott and Kenward's groupings have been applied (based on Carrott and Kenward 2001). There appears to be no clear match between the refined York ecological groups and the data from London. This most likely results from the distinctive circumstances at York and suggests that although these groupings (e.g. Carrott and Kenward 2001) are generally applicable they should be used with caution outside York. Each archaeological site is unique and, perhaps, we should not assume urban insect faunas from different locations would be identical.

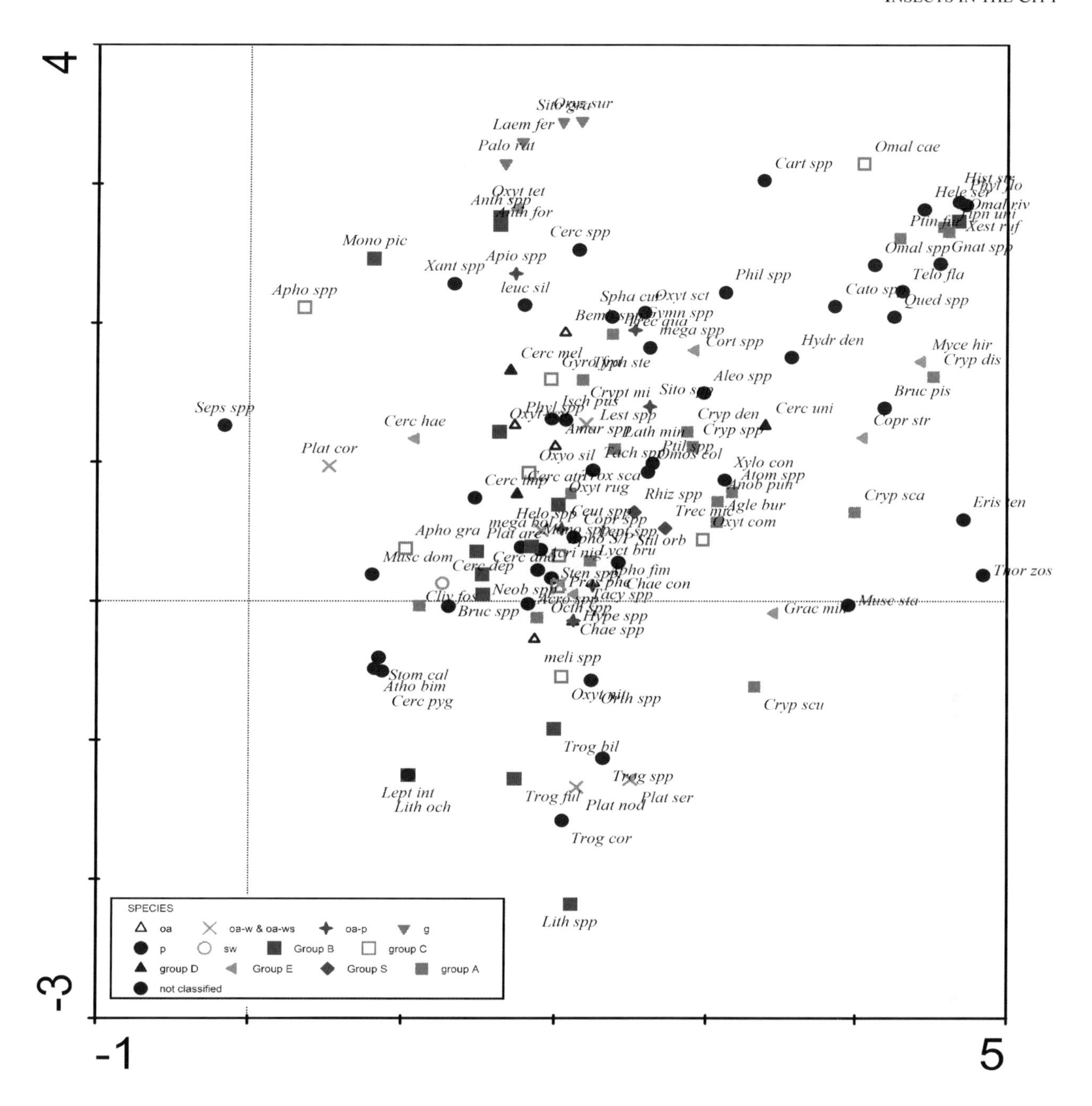

Figure 11.4. **The results for the CANOCO DCCA ordination of the London data using the ecological codes from Carrott and Kenward 2001. The membership of Carrot and Kenward's (2000) groups are outlined in Table 1.2.**

A NEW PACKAGE: CESS AND MIXED RUBBISH PITS

Intuitively, I have always felt cesspits had a distinct archaeoentomological fauna, but now with the DCCA ordination results (see Figures 11.3) of the London dataset I can suggest that it is a distinct group of taxa, which reliably occur together regardless of period. As a result, I will now discuss the cesspit fauna in detail.

The identification of a group of insects that cluster together (see the right hand side of Figure 11.3) suggests that there is a strong association of taxa which are repeatedly present in deposits that have been archaeologically identified as cesspits. This includes several 12th century AD deposits at Poultry (samples OP222, 789, 471, 621). The fills of the 15th century garderobe at Winchester Palace (i.e. samples WP5169, 5194, 5206, 5211), several of the fills of the Infirmary latrine and other cesspits from St. Mary Spital (568, 1286, 270), and various pit fills and yard deposits from Preacher's Court (PC 30, 32, 66, 25). It is notable that the features and samples, dominated by this 'cesspit' group, produced one of the few coherent groupings of samples in Figure 11.3 with cesspit and pit fill taxa clustering together at the right of the ordination. Notably these also plot out together at the right of Figure 11.4, which uses the refined York ecological groups (Carrott and Kenward 2001).

The species that cluster together in the ordination plot (see Figure 11.3) include a number of very distinctive flies. By far the most 'superabundant' of these is the small fly *Thoracochaeta zosterae*, which in many cases occurred as several hundred individuals in a sample. This is a species that Skidmore (1999) suggests is typical of archaeological cesspits. Today it is only found in accumulations of seaweed at the high water mark on the shore (Skidmore 1999). Belshaw (1989) holds that its presence probably suggests that archaeological cesspits often contained water and other substances with a highly saline nature. This most likely resulted from the inclusion of both faecal material and stale urine. Another fly commonly found as part of this fauna and in a number of pits is *Eristalis tenax*, the rat tailed maggot or the drone fly. Larvae of this species are commonly found in stagnant water containing a high concentration of decaying organic matter and faecal material (K.G.V. Smith 1973, 1989). The latrine fly, *Fannia scalaris*, which was found in the pits at Ball Wharf and Preacher's Court, has larvae that come equipped with a series of air filled spikes that enable them to float on the water's surface. The limited numbers of individuals of this species meant that it was originally excluded from the statistical analysis, but when included, it does plot out into this group. *Telomerina flavipes, Heleomyza serrata* and *Hydrotaea dentipes* also favour similar 'cesspit' environments, along with carrion, and this last species probably is also a predator of fly larvae commonly observed in these deposits. The opportunity to predate on large numbers of fly larvae probably explains the presence of the histerid 'pill beetles', such as the various species of *Hister* and *Gnathoncus,* and the large numbers of *Quedius* 'rove beetle' commonly encountered in these deposits.

Other species seem to contradict this. Several members of Kenward's 'house fauna' (Hall and Kenward 1990) were present; such as the two ptinid 'spider beetles', *Tipnus unicolor* and *Ptinus fur*, along with a number of species commonly associated with relatively dry and mouldering materials. Certainly mouldy conditions are evident due to the recovery of various *Omalium* species, *Mycetaea hirta* and several species of *Cryptophagus*, all of which are mould-feeders. How can we explain this odd association of taxa from clearly very foul circumstances with those from the relatively more pleasant, dry materials?

There are several explanations. This does return us to an old, arcane and rather amiable disagreement amongst archaeoentomologists. This started with Peter Osborne's work on the Worcester barrel latrine (Osborne 1983). He found that the fill of the barrel produced a fauna of beetles essentially similar to the 'cesspit fauna' produced in the plots presented above (see Figures 11.3 and 11.4). Osborne felt that there were two aspects of this fauna that needed further explanation and experimentation. First, what is this strange fauna doing in features that are suspected to be cesspits? Can we find a modern analogue fauna? Second, many of these pits, and indeed the CANOCO plots and faunas presented in this book, usually contain species such as the 'grain pests' and the *Bruchus* 'pea weevils' all of which are associated with stored food products.

Osborne explored the first issue by taking samples from real life. At one point Peter lived in the Wyre forest and did not have indoor plumbing. The 'necessary' was done into a 'gazunder' and this was emptied into a pit in the garden. This modern-day cesspit seemed to present Peter with the ideal modern analogue for the archaeological record and so he duly collected a sample and paraffin floated its insect fauna. (One 'urban myth' related to me was that it was quite obvious what Peter was up to in the laboratory, since semi-decayed toilet paper was clearly visible in the flot). The resulting insect fauna that resulted beautifully matched that produced from the Worcester barrel latrine.

What about the presence of pests of stored products? In Chapter 5, I mentioned Osborne's other great contribution to science; the consumption and subsequent recovery of grain beetles. The aim of this experiment was to prove that insect fragments could survive a trip through the human digestive track and enter cesspits via this route, rather than only being dumped straight into the pit (possibly within rubbish). What Osborne (1983) suggests is that the majority of the insects present in these features enter the deposits within the actual cess or intentionally live in the pit itself. The insect fauna exists as a whole rather than a disparate collection of mixed remains from various dumping events or accidentally trapped insects. On a number of occasions, usually whilst referring to one of my reports, Harry Kenward (pers. comm.) has suggested that that this may be a mistake. He is not arguing that these faunas do not represent cesspits in part but rather the material must come from a number of different sources. In particular, the 'house fauna' species probably came from domestic waste and rubbish that entered the deposit either through deliberate deposition or by accident. I agree that the story is not quite so clear-cut. It would be a mistake to see cesspits as that and nothing else. Indeed many of the pits from London discussed above, judging from the artefacts, animal bone and plant macrofossils found in them, contain both food waste and urban rubbish. I think we are wise to consider such features as a continuum with just household rubbish at one end and pure cess at the other with most pits producing assemblages representing something in between. If I am correct in this; however, it would mean that simply collecting one sample per cesspit may not be good method and a system of column sampling may be more appropriate.

We also have to consider that these pits were emptied, since the contents of cesspits were useful for fertilizer on fields or levelling for redevelopment. Furthermore, I do not find myself completely objecting to the idea that insects such as the spider beetles can live down a cesspit. Having stuck my head down a few cesspits in Northern Greece and on various archaeological excavations in the UK, as well as having talked to various 'deep greens' about 'natural toilets', the key thing in the healthy

	London (total count)	Coppergate (presence)
Thoracochaeta zosterae	1594	19
Enicmus minutus (Group)	748	340
Sepsis spp.	536	106
Cercyon analis	511	288
cf. *Telomerina flavipes*	484	-
Anobium punctatum	371	264
Philonthus spp.	364	251
Musca domestica	355	84
Sitophilus granarius	347	5
Laemophloeus ferrugineus	302	1
Aleocharinidae	287	360
Platystethus cornutus	286	123
Tipnus unicolor	286	9
Aphodius granarius	278	98
Oryzaephilus surinamensis	276	-
Oxytelus sculptus	255	237
Quedius spp.	251	36
Cercyon spp.	248	-
Aglenus brunneus	230	237
Atomaria spp.	219	268
Trogophloeus spp.	217	58
Cryptophagus spp.	214	299
Ptinus fur	198	175
Bruchus pisorum	198	-
Xylodromus concinnus	197	270
Ischiolepta cf. pussila	196	-
Cercyon atricapillus	188	148
Trogophloeus bilineatus	182	182
Corticaria/ corticarina spp.	174	202
Mycetaea hirta	164	92
Oxytelus spp.	160	-
Omalium rivulare	156	174
Cryptophagus ?scutellatus	145	146
Bruchus spp.	136	30
Oxytelus nitidulus	126	270
Xantholinus spp.	125	14
Sphaerocera curvipes	125	2
Neobisnus spp.	124	324
Oxytelus scupturatus	120	54
Lithocharis spp.	117	26
Aphodius spp.	117	263
Anthicus spp.	105	-
Oxytelus rugosus	101	257
Apion spp.	101	253
Acritus nigricornis	97	179
Stenus spp.	97	236
Copromyzinae	96	6
Hister striola	95	-
Cercyon haemorrhoidalis	93	171
Monotoma spp.	85	117
Trox scaber	85	242

Figure 11.5. Species of insects from the archaeological material from London arranged by rank order of occurrence compared to the number of samples in which they occur at Coppergate, York (Kenward and Hall 1995)

management of odour from cesspits seems to be to keep the accumulating waste matter reasonably dry and when things get 'a bit ripe' then dry matter, earth or lime is added to form a cleansing layer before accumulating the next layer of cess. Such 'cleansing layers' may well act to introduce unrelated insects, but also may form a suitable habitat for a range of species of insects, possibly several members of the house fauna, even if only for a short period.

So perhaps we should think about this blend of species repeatedly recovered from these features as an 'indicator group' (*sensu* Kenward and Hall 1997) for cesspits, at least to some extent. The archaeology of these features (notably the 'layering of deposits' using ash and lime), the artefacts present (the glass urinals from Winchester Palace being a case in point), the nature of the animal bone and plant macrofossils recovered, along with the insect remains should perhaps be used together to form an 'indicator package' for this type of cesspit deposit.

COMPARISON TO COPPERGATE, YORK

The insect faunas from the 9th century site 16–22 Coppergate, York (Kenward and Hall 1995) probably form the largest and most comparable insect faunas with those examined from London. The statistical analysis presented above clearly has demonstrated that not only is there a high degree of correspondence between the general faunas from London and York, but also that many of the 'ecological groups' identified by Kenward (1978, 1982; Hall and Kenward 1990; Kenward and Hall 1995) from Coppergate are valid elsewhere. This certainly suggests that in terms of archaeological deposits and depositional process the interpretation of these groups are comparable between sites. It also suggests that many of these 'groupings' probably represent real insect communities in the past. However, the DCCA analysis does clearly identify a few specific differences between the insect faunas from London and those from Coppergate.

This is clearly illustrated in Figure 11.5. Here I have selected the 51 most commonly recovered taxa in the London dataset and have ranked them in terms of the combined total number of individuals for the entire London dataset. I also have included Kenward's data (Kenward and Hall 1995) concerning the occurrence of species at Coppergate in the final column. Unfortunately, the information available from Coppergate is not based on numerical counts but rather on the number of contexts that a particular species occurs in. Despite this there are clear similarities and some striking differences.

Kenward's 'house fauna' (*Xylodromus concinnus, Anobium punctatum, Ptinus fur, Tipnus unicolor, Atomaria* spp., *Cryptophagus scutellatus, Cryptophagus* spp., *Lathridius minutus Mycetaea hirta and Aglenus brunneus*) clearly are abundant at both sites. Various members of Kenward's 'rt' and 'rf' groups also commonly occur at both sites (e.g. *Cercyon analis, C. atricapillus, Acritus nigricornis,* Histeridae, *Oxytelus* species, *Trogophloeus* species, *Platystethus* species,

Lithocharis species, *Neobisnus* spp., *Philonthus* spp., *Aphodius granarius* and *Trox scaber*). The flies *Musca domestica, Ischiolepta cf. pussila* and *Sphaerocera curvipes* most likely should be included in the Kenward 'rf' and 'rt' groupings as well. Certainly this seems to be indicated by the DCCA statistical analysis of the London dataset (see). It is not surprising that these two urban insect assemblages are essentially similar. Much of the nature of human behaviour, types of deposits present and the process of site formation will be quite similar not only between these two sites, but also at most urban archaeological sites.

Where there is clear variation, this can be explained by differences between the two sites in terms of either date or the types of context sampled. One obvious difference is the domination of grain pests in the material from London (Figure 11.5) and there almost total absence at Anglo-Scandinavian York. This is certainly due to the dominance of grain pests in Roman period London. This difference is clearly explained by the fact that the material from London dates to a wide range of periods; whereas the material from Coppergate only dates to a single period (the 9th century). If we expanded this exercise to include the data from the Roman periods at York (i.e. Kenward and Williams 1979; Hall *et al.* 1980; Kenward *et al.* 1986; Hall and Kenward 1990) we also would find that the grain pests were dominant. It is clear, both at York and in London, that the occurrence of grain pests and the difficulties with the storage of grain that their presence indicates are a phenomenon which is essentially limited to the Roman period.

When I discussed the statistical results earlier in this chapter, I outlined a strong grouping of both insects and flies that are associated with cesspits. These species all occur in large numbers in the data from London but are more or less absent from the Coppergate site (e.g. *Thoracochaeta zosterae,* cf. *Telomerina flavipes, and Bruchus pisorum*). This difference between the data from London and that from Coppergate probably relates to the fact that cesspits are a dominant feature of Medieval deposits from the London sites. In fact, it is notable that there appears to be a marked difference between the insect faunas from the Roman period at London and those from the Medieval period. There are few cesspit faunas from the Roman period.

At first I thought that this might be due to the fact that most floor and building deposits from Medieval archaeological sites in London are not waterlogged. Waterlogged deposits from the Medieval period tend to be pits that have been cut down into the underlying Roman layers, if not deeper. The overabundance of Medieval cesspit deposits is, in fact, a depositional problem. Although this explains why we have so many waterlogged Medieval cesspits, this does not resolve why they are entirely absent from the Roman period. It is the classic *Star Trek* conundrum: where did they go to the loo? There is no reason to presume that 'doing the necessary' never occurred in Roman London, and we certainly know from a number of multi-seater loos elsewhere in the Empire (e.g. Halsteads on Hadrian's Wall and at Douga in Tunisia, for example) that 'facilities' did exist. However, how cess was disposed of in Roman London remains a mystery.

One explanation may be that much of the settlement waste, rubbish, and presumably cess, was centrally collected and disposed of into disused quarry pits, road revetments and into the 'swamps' and diggings along the Walbrook. It might be an exaggeration to see this as some form of centralised rubbish collection/ disposal, but it may suggest that this activity was less 'home-centred' than it was in later periods. Oddly enough, the rather unglamorous question as to what happened to Roman excrement is an aspect of the past that seriously needs to be explored in much more detail in London (and indeed elsewhere in the UK).

CHAPTER 12: URBAN INSECTS: WHERE TO NOW? THE FUTURE OF ARCHAEOENTOMOLOGY IN LONDON

This survey has clearly shown some of the strengths and weaknesses of urban archaeoentomology in London. Figure 12.1 outlines the number of sites with archaeoentomological results by archaeological period.

Period	Number of sites
Lower Palaeolithic (anything pre 250,000/ 200,000)	1
Middle Palaeolithic (250,000/ 200,000 – 40,000)	1
Upper Palaeolithic (40,000 – 10,500)	3
Late Glacial/ Early Post-glacial (10,500 – 9500)	0
Mesolithic (9500 – 4500)	2
Neolithic (4500 – 2000)	3
Bronze Age (2000 – 700)	6
Iron Age (700 – 50)	0
Roman (50 – 410)	3
Early Saxon (410 – 900)	0
Late Saxon/ Early Norman (900 – 1200)	3
Medieval (1200 – 1500)	5
Post-Dissolution (1500 +)	6

Figure 12.1. Number of sites containing insect remains in Greater London by Archaeological period

Despite the length of Chapter 2, which dealt with the Palaeolithic and its insect faunas, the distribution of sites across such a long period of time is extremely patchy. Five sites are not sufficient to characterize 300,000 years of London's past. The Lower and Middle Palaeolithic (*ca.* 350,000 – 40,000 BP) are represented by only two sites, Nightingale Farm and Trafalgar Square. The latter has not been published or fully reported. To some extent the lack of sites for this period is not a surprise. Sites of this date are rare nationally and we should not expect the Thames Basin to contain many. The geological record in the Thames Valley clearly shows a considerable degree of re-working of floodplain deposits by the Thames in the Upper Palaeolithic and the Holocene (Sidell *et al.* 2000, 2002). Many deposits from these earlier dates are likely to have been destroyed as a result.

The situation seems better for the Upper Palaeolithic (40,000 – 10,500 BP). There are three sites, all of which are published (Kempton Park, Isleworth, South Kensington). However, these all date from around 40,000 – 35,000 BP. This is the Middle/ Upper Palaeolithic transition and actually deciding which period these sites should be slotted into is not straightforward. The situation at Kempton Park, where the youngest deposits from this period were sampled is no easier. The work at this site was complicated further by the coarse-sampling interval that was used on site (Gibbard *et al.* 1982).

What is striking is the complete lack of insect faunas between *ca.* 30,000 BP until *ca.* 5000 BP. This is not an insubstantial period either in terms of time or archaeology. The lack of Late Glacial/ Early Post-Glacial faunas is particularly surprising when held against other British river systems such as that of the Trent (Greenwood and Smith 2005; Knight and Howard 2005) or indeed the Upper Thames Valley north of the Goring Gap (Robinson 1978, 1979, 1991, 1993, 2000a).

A number of sites in the Lower Thames region dating to this period have received considerable palaeoenvironmental attention and did produce waterlogged deposits, but insect remains were not sampled or analysed. To be fair, many of these sites were not excavated as open sections or were in places where access was limited. Collecting 10 kg samples in such circumstances can be difficult; nevertheless, this still represents a lost opportunity. It is of vital importance to our understanding of this period in the Thames Valley that in future when deposits of this date are encountered insect analysis is included in any research or rescue programme that occurs.

Things are actually no less depressing when we look at the extent of archaeoentomological work undertaken in the Mesolithic and Neolithic periods. Robinson's (1991, 2000a) work at Runnymede Bridge helps but the faunas are from a series of palaeochannel features often with considerable temporal *hiatūs* between sampled deposits. Though this is not in itself a problem, since such deposits can be 'stitched' together to form complete sequences (e.g. Smith *et al.* 2005), this still only represents a partial record. Though there is a more complete sequence from West Heath Spa (Girling and Greig 1977; Girling 1989a) it is poorly dated, only relying on the presence of the elm decline in the pollen spectra. This essentially leaves our knowledge of the insect fauna of the Thames Basin during the Mesolithic and Neolithic largely incomplete and certainly imprecisely dated. Another factor is that only riverine and 'high woodland' landscapes are represented. We have no representation of insect faunas from the other landscapes that must have existed in this region. For example, how did the interplay of estuary, salt marsh, fresh water, fen dry land and woodland effect the development of insect faunas in the lower Thames Valley to the east of the present city? Also it is very difficult to understand the interplay between the developments of wildwood, 'cultured' woodland, land under farming and/or pasture regimes in the London area after 1000 BC without understanding the diversity of the indigenous insect fauna at this time, or previously (i.e. prior to human modification).

Similar problems exist with the insect assemblages from Bronze Age deposits. These are essentially limited to a number of trackways and similar features that cross over raised bog and fen to the east of modern London. This lack of diversity in the insect faunas, and the locations they come from, is deeply worrying. Finally, there are no representative insect faunas for the Iron Age as a whole (though this situation, at least in terms of the area around Heathrow, is improving).

Once we arrive at the Roman period things improve. There are three Roman sites where extensive sampling and analysis of insect faunas has occurred. As Chapter 5 showed, we are beginning to have a good handle on the nature of the urban environment, and the insect faunas present from the earlier Roman periods. However, again there are some areas of concern where further research is urgently needed. In terms of 'average' urban deposits from this date the best material comes from Poultry (see Chapter 5) and the later periods from the sites in the Upper Walbrook Valley (de Moulins 1990) (I do not think we should regard the deposits and activities associated with the Guildhall amphitheatre as 'typical'). However, both Poultry and Guildhall are not actually located in the 'heart' of Roman London, which was centred on the area around modern Cornhill. I suspect that settlement in this area would be essentially the same as that suggested at Poultry, but it would be nice to have the opportunity to put this to the test. Another clear gap in our knowledge is that deposits from the Late Roman period also are comparatively rare. Aside from the early deposits at Copthall Avenue in the Upper Walbrook (de Moulins 1990), there are no Roman insect faunas from 'outside' of the town itself and as a result we have little idea of the non-urban fauna and landscape in the wider area. This is also a common feature of Roman archaeology in London in general (Rackham and Sidell 2000).

One period where there are no insect faunas recovered at all is the early Saxon (*ca.* AD 400 – 900); a truly dark age archaeoentomologically at least. In some ways this is clearly understandable. The settlement of *Lundenwic*, centred in the area around modern Covent Garden, lies in a part of London that does not seem to include waterlogged deposits. If waterlogged Early Saxon deposits are encountered sampling for insect remains is paramount. The coverage of later Saxon and Norman urban deposits is actually quite good, although the lack of deposits of a more 'natural' origin away from the town is marked and clearly should be addressed. Again this is an issue for the archaeology of London in this period in general (Rackham and Sidell 2000).

This latter problem also occurs with the Medieval and Post-Dissolution faunas, with the vast majority of the insect faunas being clearly urban in nature. Again charges of being unrepresentative could be made. There is a distinct bias towards the monastic rather than the domestic in terms of site type and location. Physically, these consist of a ring of sites, often outside of the city wall, surrounding the urban heart of London at this time. However, to be fair, cesspits are cesspits and dumps of stabling manure are dumps of stabling manure, regardless of situation. Presumably similar deposits were common in Central London at this time, but this needs to be proven.

Another problem with the present coverage of urban insect faunas is that some types of deposit and sample are more common than others. Figure 12.2 shows the

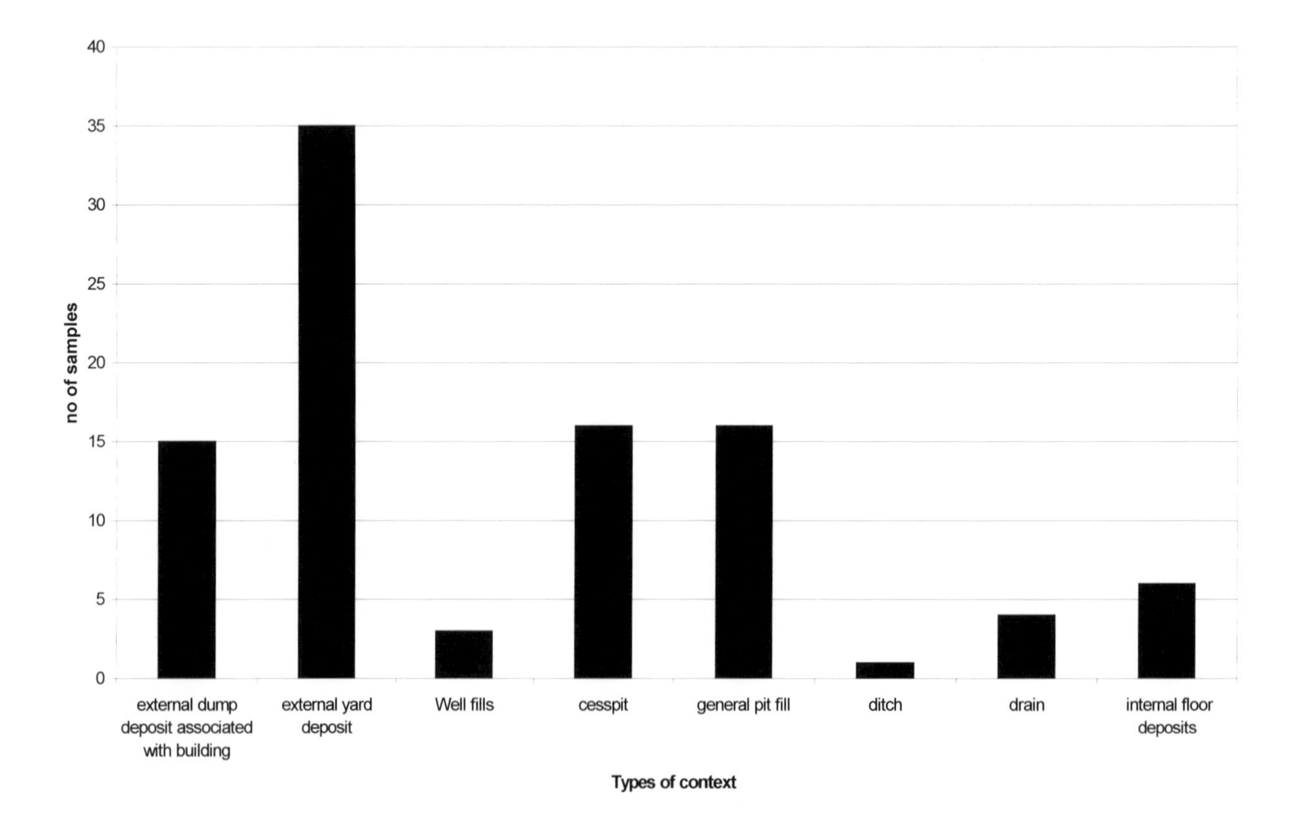

Figure 12.2. The numbers of samples from different context types sampled from urban London in all periods

numbers of the types of samples analysed as part of this survey.

The vast majority fall into those horrible archaeological catchalls of 'external dump deposit' or 'external yard deposit'. As we saw at the Guildhall site these often are not deposits that could be regarded as *in situ* or undisturbed. This is in clear contrast to the situation at York, certainly at the Coppergate site, where rather more of the material seems to have been undisturbed (Kenward and Hall 1995). A good proportion of the material sampled at Coppergate came from undisturbed internal house floors. House, or rather building, floors are relatively rare in the material from London; limited to a small number from the building at Roman Copthall Avenue (de Moulins 1990) and several from 11th and 12th century Guildhall. As has been noted previously, those from the Guildhall are probably re-deposited and disturbed.

Finally, and perhaps more worrying in the long term, Scott Elias and I seem to be the only archaeoentomologists working on such deposits from London. What happens if we both get hit by a bus on the way home tonight? What happens if the Museum of London Archaeology Service loses their touching faith in me? There are other archaeological organisations, societies and groups, which regularly work in the Lower Thames Valley and in Urban London, but strangely, they do not seem to require any assistance with archaeoentomological sampling and analysis.

THE FUTURE OF ARCHAEOENTOMOLOGY IN GENERAL

Personally, I am a poor predictor of the future but here are my current views on the subject, for what they are worth. In the main, particularly in terms of urban deposits, I think we will continue to develop our existing set of tools. The 'indicator package' approach, where archaeoentomological results are combined with results from a range of other analyses (e.g. Kenward and Hall 1997), seems to be successful and produces results that appear to be useful across a range of archaeological sites. It also seems to bring archaeoentomology into a wider range of archaeological discussions and interpretations than previously. I also feel that given the range of data that could be included in the various 'packages' some very strong inferences about the nature of urban life can be drawn. I suppose the main danger is that the various packages come to be used mechanically without questioning their strength on a site-by-site, or indeed context-by-context, basis (like the 'Pavlovian response' to indicator groups mentioned in Chapter 6). Finally, the use of multivariate and other 'high level' statistics to identify consistent groups of insects from the archaeological record does seem to be producing real results, which can facilitate interpretation and test assumptions.

Working on modern analogue faunas and sub-fossil faunas to help us understand depositional problems and what our faunas actually mean is helpful. In the past my own work in this area has at least had value in demonstrating some of our limitations (Smith 1996a, 1998, 2000a, Smith *et al.* 1999, 2003). In a similar vein, Nicki Whitehouse (Queen's University, Belfast) and I have developed a project where we are collecting sediment from modern ponds in a variety of grazed and ungrazed woodlands and grasslands in order to establish just what the 10% open ground beetle fauna in early Holocene samples means in terms of the actual canopy cover and how this applies to the 'Vera hypothesis' (see discussion on woodland clearance in Chapter 4).

BEYOND THE BEETLES AND 'INTERPRETIVE ARCHAEOLOGY'

This book has shown that insect remains in archaeology can be used to answer a number of questions about the past. However, one clear failing is that many of these are essentially descriptive rather than 'explanatory' or interpretive. In the world of modern academia, archaeological work such as that described in this book is considered to be very out of date, functionalist and to have no real involvement with the process of modern theory building or 'mainstream post-processual' archaeology or, indeed, any of the issues of major significance (J. Thomas 1990, 1999; Van de Noort and O' Sullivan 2006). Environmental Archaeology is seen as being predominantly concerned with the collection of empirical data, description, economic reconstruction and climatic and environmental reconstruction (J. Thomas 1990; Wilkinson and Stevens 2003). It is often seen in the 'wider archaeological' community as the last bastion of the 'old archaeology'; only concerned with the process of data collection and scientific analysis, tending towards naive and simplistic mechanistic explanations of human change and behaviour (Thomas 1990; Tilly 1994; Wilkinson and Stevens 2003).

In addition, environmental archaeology often tends to work on a very local and detailed level, such as that of the context, site or the community rather than link into larger regional and social research questions (Whittle 2000, 2; Van de Noort and O'Sullavan 2006). Environmental archaeologists spend a lot of time worrying over how archaeological material was obtained, processed, used and disposed of. We remain very interested in how to identify such materials and activities. Our attention to such details can make our approach appear mechanical and almost dehumanising. Indeed, this book can be seen to be following such an approach. It really *has* been about the beetles and the deposits in which they were found. It rarely touches on the people who formed the archaeology and why they may have chosen to do so. More importantly, perhaps, it fails to address what such activities could achieve and what it meant to people. This approach to the past has been attacked as being 'processual' (i.e. concerned with the process and nature of the archaeological record, itself, rather than informing us about human life and society and how people experienced and understood the world around them).

For an example of how this may work on the ground we should look at Alistair Whittle's introduction to *'Plants in the Neolithic and Beyond'* (Fairbairn 2000). This collection of papers could be seen as a deliberate attempt by a range of environmental archaeologists working on the economy and landscape of the Neolithic to engage with larger archaeological questions. Two themes mark these papers. One is the possible role of clearance and grazing in the Neolithic landscape and how this might relate to its use by people (Allen 2000; Robinson 2000a; Brown 2000). The other relates to issues raised by Thomas' (1991) suggestion that Neolithic people may not have been as dependent on farming and cereals as previously thought. Central to this is the debate between G. Jones (2000) and Robinson (2000c) on the interpretation of charred and waterlogged plant remains from the Neolithic. For many years I have regarded this as one of the most stimulating texts on my bookshelf and have used this book with students to show the extent to which the kind of work I do can help inform current archaeological debate. Perhaps I should have read the introduction paper by Alistair Whittle (2000) with more care, for I seem to have missed the point (or perhaps even the fast departing boat of academic debate). Whittle's paper is a gentle reminder, and slight prod, to look beyond the associated issues to the nature of the main arguments being made at present amongst Neolithic archaeologists.

He suggests that much of the history of research into plant use in the Neolithic is centred on the large-scale and 'totalising' issues such as the introduction, use and processing of domesticated plants (Whittle 2000, 1). This is an approach to the past, which like a lot of environmental archaeology, focuses on overarching and cross cultural explanation, or what has come to be called the 'meta-narrative' (van de Noort and O'Sullavan 2006). Equally, there is tendency, because we regard the work we do as important to, consciously or unconsciously, slip into environmental determinism to explain the past.

Perhaps most damming he notes *'a lot of emphasis in these papers [is] on context and setting in a broad and rather neutral sense'* (Whittle 2000, 2). This is fundamentally where the difference in approach between environmental archaeologists and 'theoretical' or 'interpretive' archaeologists lies. Many of my fellow environmental archaeologists, and indeed archaeologists, tend to see what we do as providing a stage canvas against which humans act and archaeological sites can be reconstructed (see Caseldine *et al.* 2008 for a criticism of this). Many 'interpretive archaeologists' hold that this is a rather simplistic view of the past and the past actions of people. People experienced their landscape as part of social action and would have given it meaning (Tilley 1994, 2003; Bradley 2000). The crops they grew and the animals they tended would have had specific social meanings and their use and the way in which their remains would have been disposed of would have been imbued with specific meanings (Hill 1995; J. Thomas 1991). They would all have been part of the world of 'symbols' that helped people understand the way the world worked. It is trying to 'experience' or at least identify these patterns and give them possible social meaning that is at the heart of 'interpretive archaeology'. Why can't environmental archaeology do this?

LOOK! MY BEETLES (AND I) CAN DO IT AS WELL!

Can insect remains be used to play the same intellectual games? Let's look at two issues of archaeological explanation common in 'theory-based' archaeology at present.

Intervisibility and environmental siting of Neolithic monuments

'Phenomology of landscape' is a technique, and way of viewing landscape, proposed by Chris Tilley (1994; 2003). In this approach the relationship between Neolithic monuments and their landscape is seen as central to allowing us to understand the society that builds them. It is held that these societies 'encultured the landscape' and this structuring would leave clues to meaning that we can 'experience' and attempt to understand. Key to this is the idea that the movement and pathways through landscape may have had specific histories, use and meaning. Understanding how these experiences were used and changed through time is one way of accessing past social action. The approach and fieldwork itself has been criticized extensively (Fleming 1999; Bradley 2000; Wilkinson and Stevens 2003) but understanding the location and place of monuments in the landscape, and how this may have related to social practice and action, has now become a central theme in mainstream Neolithic archaeology (e.g. Bradley 1997; Whittle and Pollard 1999; Bradley 2000; Exon *et al.* 2000; Scarre 2002; Whittle 2002; Pollard 2004).

One problem that is raised is that phenomology depends on the 'intervisability' of the monuments, and features of the landscape that surround them. This is a factor that may have been considerably complicated by tree cover (Fleming 1999; Evans 2003; Cummings and Whittle 2003; Davies *et al.* 2005). Attempts to come to grips with this problem range from ignoring it (Tilley 1994), suggesting that it is knowledge of the location of landscape features rather than visibility that matters (Whittle and Pollard 1999; Cummings and Whittle 2003; Pollard 2004; Tilley 2003) on to arguing that where monuments are common, the surrounding area must have been substantially cleared or under secondary woodland (J. Thomas 1991; Allen 1997; Austin 2000; Evans 2003; Cummings and Whittle 2003). Several people have suggested that 'impermanent' trees, groves, forest and clearings may have had as much meaning as the 'hard' features of the landscape and this needs to be considered (Fleming 1999; Whittle and Pollard 1999; Evans *et al.* 1999; Austin 2000; Pollard 2000; Bradley 2000; Pollard 2004; Davies *et al.* 2005). In addition there often is a plea to see this heavily wooded landscape as not just 'ecological background'; but to also consider what this forested landscape *meant* to people, how it was perceived and how it could be used to symbolise relationships in society and with the spirit world (J. Thomas 1991; Evans *et al.* 1999; Whittle and Pollard 1999; Bradley 2000, Pollard 2000; Austin 2000; Pollard 2004).

A common thread through much of this discussion is the assumption that Neolithic Britain lay under dense 'primeval' woodland where clearings were the exception and pathways through the forest were constrained. There have even been suggestions that early Neolithic monuments such as causeway camps may have marked the edges of existing clearings, perhaps occurring where paths crossed through dense woodland (J. Thomas 1991; Tilley 1994; Whittle and Pollard 1999; Austin 2000; Pollard 2004). By extension there is often the hint that later henges, though set into open landscapes, may symbolically function in the same way as the earlier clearings (J. Thomas 1991; Austin 2000).

In Chapter 3, I discussed the kind of landscape, or rather forestscape or treescapes (i.e. Pollard 2004), which the 'Vera hypothesis' might suggest existed at this time and the role that insect remains may play in establishing the validity of this hypothesis. If Vera is right we may need to rethink many of the relationships between monuments and the landscape and how this may symbolise social action. Vera's landscape is not dominated by dense woodland but consists of groves of woodland and scrub divided by extensive and long-lived clearings. The landscape reconstruction therefore opens out to allow visibility without the clumsy need for human intervention. However, other problems are raised. If clearings are common why give one a specific meaning over another? If pathways between places are not constrained by the surrounding woodland why should any pathway across the landscape become more permanent or have a deeper meaning than any other?

Perhaps we should also stop thinking of people acting in 'tamed' clearings in an otherwise wild landscape (e.g. J. Thomas 1991; Whittle and Pollard 1999; Pollard 2004). Perhaps the monuments and the actions that occurred in them may have been sited in, or in relation to, the groves in Vera's proposed Neolithic landscape. Instead of seeing the later monuments such as the henges and woodhenges as symbolising clearings, should we rather see them as symbolic groves (this seems especially obvious for the woodhenges with their dense 'stands' of vertical timber). It may be the hidden and furtive nature of actions conducted in symbolic 'sacred groves' and the obscuring of action as it takes place in a relatively clear landscape that is of importance here. Indeed for a discussion of the use of 'sacred groves' in a range of societies see Bradley (2000).

'Structured deposition'

Another recent trend in archaeology is to concentrate on the identification of 'structured deposition'. This is an extension of the idea that 'ritual deposits' can be identified in the archaeological record. Often the definition of a 'ritual deposit' is one that we cannot understand, involving features and activities that are associated with obvious 'ritual sites'. For example, if you are digging a chambered tomb you might expect to find ritual deposits; if you are digging a medieval tannery you might not. 'Structured deposition' is an attempt to try to extend this search for 'significant deposits' away from the obvious ritual and religious contexts to the more mundane locations and activities. The term was first coined by Richards and Thomas (1984) and subsequently expanded on by J.D. Hill (1995). Hill (1995) investigated pit fills from a number of Iron Age settlements in Wessex. He found that a third of the deposits studied appeared to result from frequent, formalised and repetitive actions. In particular, there often were recurrent patterns in terms of the associations between the types of material present and the contexts in which they were found. For example the deposition of human remains, whole or jointed parts of domestic animals (associated bone groups or ABG), a range of small finds, particularly wild animals and various classes of pottery all seemed to be linked. The key thing to grasp here is that the pits studied were not located on 'ritual' or religious sites, but were on 'mundane' settlements. From Hill's work it was clear that many settlement deposits were as much a part of an active ritual and religions life as those found solely at 'sacred sites'. This all fitted well with one of the key components of 'interpretative archaeology'. This is the idea that all human action in the past involved some degree of ritual practice and inherently had the potential to act as 'symbols' expressing the structure and nature of the society present at that time (e.g. Shanks and Tilley 1982; 1987).

Recently, the search for examples of 'structured deposition' and the extension of 'ritual activity' in to the world of the 'every day life' and settlement archaeology has produced a number of other examples. These range from the 'digging pits' of the Early Neolithic (J. Thomas 1991; Pollard 1999), foundation shafts in the early Romano-British period (Woodward and Woodward 2004), and the reinterpretation of certain features of 'wetland archaeology' (van de Noort and O' Sullivan 2006).

I find this to be an attractive concept. For one thing its application is essentially similar to the 'indicator package' concept put forward by Kenward and Hall (1997) when considering the interpretation of the bioarchaeology of urban deposits. It is the use of a recognised set of criteria to define and detect an action in the archaeological past (I apologise for being 'reductionist' here (e.g. Hill 1995, 97) but essentially it is true). In practice, I very rarely get to work on material associated from 'ritual sites' and the idea that I can investigate socially significant materials and contexts even from 'boring old urban stuff' is pleasing.

However, I do have some problems, particularly in the way that language has been used. Both 'structured deposition' and the labelling of deposits as having 'significance' are loaded terms. They suggest that less prosaic deposits such as rubbish pits or cesspits have no significance or are, indeed, 'unstructured'. By extension this suggests that they have less value. Also common to this frame of mind is the idea that the 'mundane' activities and behaviour of an archaeological site are of less interest or importance to the 'modern' archaeologist.

I have recently attended a number of academic conferences where questions concerning the economic, industrial and depositional histories of material and archaeology are dismissed with the mealy mouthed phrase 'these are important questions to ask of the material but….'. I begin to fear that there is a slippery slope into the situation where all deposits are regarded as potentially 'structured' and, intimately involved with ritual and, therefore, statements of social structures. I recently heard an academic paper where someone, who should have known better, proposed that ALL deposits from the Iron Age that contained charred plant remains must have had some ritual nature. Let us return to my cesspits as an example of this fallacy. I am fairly certain that the everyday use of cesspits does not normally involve prayer or ritual behaviour but the pits themselves often are highly structured and deliberate in terms of fills and contents. However, that said, given the apparent 'dangerous' nature of the material in cesspits I am happy to concede the point that these pits would be viewed as 'unclean' and that the digging and closing of such pits may have entailed 'purity' rituals associated with these activities. It is all a matter of context.

I think a fairer way of seeing the archaeological record, and the place of environmental archaeology in 'interpretative archaeology', can be seen in the ideas put forward by John Evans (2003). He urged that environmental archaeology must 'key in' with mainstream thought. He suggests, borrowing from the ideas of Leroi-Gourhan (1993 [1964]), that all archaeological materials need to be seen as part of a '*chaîne opératoire*' where production, use, disposal and social meaning are all interlinked. You cannot separate the meaning of one from the other. You must therefore understand the whole 'past life' of the archaeology you study, even if at one stage in its life it takes on a strong 'ritual significance'. This is an idea which has a history of use in lithic studies (e.g. Conneller 2008) and a similar line of argument has recently been seen in landscape archaeology were the idea of 'biography' of a landscape has become current (Pollard and Reynolds 2002; Darvill 2006).

How does this involve my beetles? Well they are supremely good at three things; identifying what materials are present, the condition of those materials and the degree of complexity of deposition that has taken place. These are all factors that can be helpful in terms of identifying 'structured deposition'. For example, let us say we have a waterlogged deposit that is identified as being the result of 'structured deposition' and that my 'cesspit fauna' is present. This might suggest that one of the most potent and symbolic of substances has been used in the creation of this deposit. This leads me full circle in terms of my academic life. The first book I read when I started my undergraduate degree was *Purity and Danger* by Mary Douglas (1966) where the role of faecal material and other spiritually dangerous materials in religious and ritual acts was examined.

So why don't I play these games more often? To be fair I am probably guilty of many of the crimes suggested by J. Thomas (1990) when he discussed the 'ills of environmental archaeology'. In particular, what I do is time consuming and often represents work to tight deadlines. It is perhaps no wonder that it is easier in such circumstances to fall back on the descriptive. Most of the deposits I look at, and the sites on which I work, are not particularly promising in terms of ritual deposits (though given the arguments of the 'interpretative archaeologists' related above this probably needs some consideration). Far too often, the site-based and context information I receive from the archaeologist is sketchy and incomplete. Even when full details are provided by a site director or project manager, what I receive is often factual and descriptive itself (e.g. context 2525 – 'layer'). I very rarely get told of the larger interpretive issues at the site under study or if particular deposits have 'special significance'. Often I know more about the pollen present in a deposit, than that deposit's possible interpretation when I write up my results. I frequently have the distinct impression that 'the interpretation thing' is thought to be solely in the purview of the archaeologist and not up to me. Julian Thomas (1990) suggested that *'there should simply be archaeologists who specialise in the analysis of snails, pollen and seeds'*. Personally, I agree with this point. However, we also need to be included in the 'interpretative conversation' that should occur about the nature of deposits as much as the excavator or the pottery expert. Perhaps, in addition to discussing how we interpret archaeology we also should consider how we avoid disjointed archaeology in practice, particularly as we work in a more commercial environment.

Appendix: Summary taxa list for the sites from London. Nomenclature follows that of Lucht (1979).

	Canoco codes	Ecological codes	Synanthropic codes	Total Samples analysed	Poultry PB	Poultry 1st and 2nd century	Guildhall 2nd century	Poultry 3rd century	Poultry Saxon / Norman	Guildhall Saxon /Norman	Ball wharf Saxon/ Norman	Preachers Court Medieval	St. Johns Medieval	Winchester Palace Medieval	St. Mary's Spital / Medieval
DERMAPTERA															
Forficulidae															
Forficula auricularia (L.)				15	0	1	0	0	14	0	0	0	0	0	0
MALLOPHAGA															
Damalina spp.				2	2	0	0	0	0	0	0	0	0	0	0
HEMIPTERA															
Family, genus and spp. Indet.				19	2	4	0	2	11	0	0	0	0	0	0
COLEOPTERA															
Carabidae															
Carabus granulatus L.	*Cara gra*	oa		1	0	0	0	0	0	1	0	0	0	0	0
Nebria brevicollis (F.)	*Nebr bre*	oa		4	1	0	0	0	2	0	0	1	0	0	0
Nebria spp.	*Nebr spp*	oa		1	0	0	0	0	0	0	1	0	0	0	0
Notiophilus biguttatus (F.)	*Noti big*	oa		2	0	1	0	0	0	0	1	0	0	0	0
Clivina fossor (L.)	*Cliv fos*	oa		36	3	5	1	0	0	20	5	0	1	1	0
Clivina contracta (Fourcr.)	*Cliv con*	oa		3	0	2	0	0	1	0	0	0	0	0	0
Dyschirius globosus (Hbst.)	*Dysc glo*	oa		5	4	1	0	0	0	0	0	0	0	0	0
Trechus rubens (F.)	*Trec rub*	oa		2	0	1	0	0	0	0	1	0	0	0	0
Trechus quadristriatus (Schrk)	*Trec qua*	oa		21	8	4	0	1	5	0	1	1	1	0	0
Trechus quadristriatus (Schrk)*T. obtusus* Er.	*Trec q/o*	oa		16	1	2	0	0	4	3	5	0	1	0	0
Lasiotrechus discus (F.)	*Lasi dis*	oa		1	0	1	0	0	0	0	0	0	0	0	0
Trechoblemus micros (Hbst.)	*Trec mic*	oa		16	0	0	0	4	4	2	5	0	1	0	0
Trechus spp.	*Trec spp*	oa		8	0	2	0	0	3	0	1	0	0	2	7
Tachys scutellaris Steph.	*Tach scu*	oa-c		1	0	0	0	0	0	1	0	0	0	0	0
Bembidion lampros (Hbst.)	*Bemb lam*	oa		4	3	1	0	0	0	0	0	0	0	0	0
B. nitidulum (Marsh.)	*Bemb nit*	oa		1	1	0	0	0	0	0	0	0	0	0	0
Bembidion tetracolum Say	*Bemb tet*	oa		2	0	1	0	0	0	0	1	0	0	0	0
B. assimile Gyll.	*Bemb ass*	oa		3	0	1	0	0	0	0	2	0	0	0	0
Bembidion doris (Panz.)	*Bemb dor*	oa		3	0	1	0	0	0	0	2	0	0	0	0
Bembidion obtusum Serv.	*Bemb obt*	oa		1	1	0	0	0	0	0	0	0	0	0	0
B. guttula (F.)	*Benb gut*	oa		2	0	1	0	0	0	0	1	0	0	0	0
B. lunulatum (Fourcr.)	*Bemb lun*	oa		1	1	0	0	0	0	0	0	0	0	0	0
B. spp.	*Bemb spp*	oa		41	4	8	1	2	4	5	11	0	6	0	3
Asaphidion flavipes (L.)	*Asap fla*	oa-ws		1	1	0	0	0	0	0	0	0	0	0	0
Patrobus sp.	*Patr spp*	oa		2	0	1	0	0	0	0	1	0	0	0	0
Anisodactylus binotatus (F.)	*Anis bin*	u		6	4	1	0	0	1	0	0	0	0	0	0
Harpalus aeneus (F.)	*Harp aen*	oa		1	0	0	0	0	0	0	0	0	1	0	0
Harpalus spp.	*Harp spp*	oa		11	2	4	0	0	2	1	1	0	1	0	0
Bradycellus spp.	*Brad spp*	oa		5	1	3	0	0	1	0	0	0	0	0	0
Acupalpus exiguus (Dej.)	*Acup exi*	oa		2	0	1	0	0	0	0	1	0	0	0	0
Stomis pumicatus (Panz.)	*Stom pum*	oa		3	0	0	0	0	2	0	0	1	0	0	0
Poecilus versicolor (Sturm)	*Poec ver*	oa		1	0	1	0	0	0	0	0	0	0	0	0
Pterostichus strenuus (Panz.)	*Pter stre*	oa		2	0	1	0	0	0	0	1	0	0	0	0
Pterostichus diligens (Sturm)	*Pter dil*	oa		3	3	0	0	0	0	0	0	0	0	0	0
P. anthracinus (Ill.)	*Pter ant*	oa-d		2	0	1	0	0	0	0	1	0	0	0	0
P. minor (Gyll.)	*Pter min*	ws		3	0	1	0	0	0	0	2	0	0	0	0
P. melanarius (Ill.)	*Pter mel*	oa		5	1	1	0	0	0	2	0	1	0	0	3
P. madidus (F.)	*Pter mad*	oa		7	0	0	0	0	3	0	0	1	3	0	0
Calathus fuscipes (Goeze)	*Cala fus*	oa		1	0	1	0	0	0	0	0	0	0	0	0
*Pristonychus terricola (*Hbst.)	*Pris ter*	u	ss	3	1	0	0	0	1	0	1	0	0	0	0
Agonum marginatum (L.)	*Agon mar*	oa-d		2	0	1	0	0	0	0	1	0	0	0	0
Agonum muelleri (Hbst.)	*Agon mus*	oa		1	0	1	0	0	0	0	0	0	0	0	0
Agonum sp.	*Agon spp*	oa		7	0	0	0	0	0	1	4	1	1	0	0
Platynus assimilis (Payk.)	*Plat ass*	oa		2	0	1	0	0	0	0	1	0	0	0	0
Platynus dorsalis (Pont.)	*Plat dor*	oa		4	1	0	0	1	1	0	1	0	0	0	0
Amara aenea (Geer)	*Amar aen*	oa		2	0	0	0	0	0	0	0	0	2	0	0
Amara familiaris (Duft.)	*Amar fam*	oa		1	0	0	0	0	0	1	0	0	0	0	0
Amara spp.	*Amar spp*	oa		25	6	5	0	1	1	1	2	1	6	2	1
Odacantha melanura (L.)	*Odac mel*	oa-ws		1	0	0	0	0	1	0	0	0	0	0	0
Dromius longiceps Dej.	*Drom lon*	oa		3	0	1	0	0	0	0	2	0	0	0	0
Dromius linearis (Ol.)	*Drom lin*	oa-ws		3	0	1	0	0	0	1	1	0	0	0	0
Syntomus foveatus (Fourcr.)	*Synt fov*	oa		2	1	1	0	0	0	0	0	0	0	0	0
Microlestes maurus (Sturm)	*Micr mau*	oa		1	0	0	0	1	0	0	0	0	0	0	0
Drypta dentata (Rossi)	*Dryp den*	oa		1	0	1	0	0	0	0	0	0	0	0	0
Dytiscidae															
Hydroporus spp.	*Hydr spp*	oa-w		3	1	1	0	0	0	0	0	1	0	0	0
Colymbetes fuscus L.	*Coly fus*			0	0	0	0	0	0	0	0	0	0	0	2
Agabus spp.	*Agab spp*	oa-w		1	1	0	0	0	0	0	0	0	0	0	0
Agabus bipustulatus (L.)	*Agab bip*	oa-w		1	0	0	0	0	0	0	0	0	1	0	0
Hydraenidae															
Ochthebius bicolon Germ.	*Octh bic*	oa-w		2	0	0	0	0	0	0	2	0	0	0	0
Ochthebius minimus (F.)	*Octh min*	oa-w		11	1	1	0	0	2	0	7	0	0	0	0
Ochthebius marinus (Payk.)	*Octh mar*	oa-w-c		0	0	0	0	0	0	0	0	0	0	0	0
O. spp.	*Octh spp*	oa-w		34	12	10	2	0	0	1	8	0	1	0	0

	Canoco codes	Ecological codes	Synanthropic codes	Total Samples analysed	Poultry PB	Poultry 1st and 2nd century	Guildhall 2nd century	Poultry 3rd century	Poultry Saxon / Norman	Guildhall Saxon /Norman	Ball wharf Saxon/ Norman	Preachers Court Medieval	St. Johns Medieval	Winchester Palace Medieval	St. Mary's Spital / Medieval
Limnebius spp.	*Limn spp*	oa-w		5	1	0	0	1	0	2	1	0	0	0	2
Hydrochus elongatus (Schall.)	*Hydr elo*	oa-w		4	1	0	0	0	2	0	0	0	1	0	0
Hydrochus aquaticus (L.)	*Hydr aqu*	oa-w		2	1	1	0	0	0	0	0	0	0	0	0
Helophorus porculus Bedel	*Helo por*	oa-w		1	0	0	0	0	0	0	0	0	1	0	0
Helophorus spp.	*Helo spp*	oa-w		71	10	13	2	0	0	4	20	1	20	1	25
Hydrophilidae															
Coelostoma orbiculare (F.)	*Coel orb*	oa-w		1	1	0	0	0	0	0	0	0	0	0	0
Sphaeridium lunatum F.	*Spha lun*	rf		2	0	0	0	0	0	0	0	0	2	0	0
Sphaeridium sp.	*Spha spp*	oa-rf		3	0	0	1	0	0	0	0	0	2	0	2
Cercyon depressus Steph.	*Cerc dep*	sw		29	0	5	0	0	0	12	10	0	2	0	0
Cercyon impressus (Sturm)	*Cerc imp*	rf	sf	31	4	9	0	2	0	4	0	1	11	0	0
Cercyon haemorrhoidalis (F.)	*Cerc hae*	rf	sf	93	2	0	11	0	8	28	1	3	40	0	0
Cercyon melanocephalus (L.)	*Cerc mel*	rf	sf	19	0	0	0	0	0	2	0	0	17	0	0
Cercyon unipunctatus (L.)	*Cerc uni*	rf	st	28	1	2	0	0	5	1	2	3	14	0	1
Cercyon atricapillus (Marsh.)	*Cerc atr*	rf	st	188	20	64	4	5	5	9	27	7	47	0	0
Cercyon pygmaeus (Ill.)	*Cerc pyg*	rf	st	14	0	0	0	0	0	3	0	0	1	10	0
Cercyon analis (Payk.)	*Cerc ana*	rt	sf	511	61	204	16	6	21	84	87	5	17	10	0
Cercyon convexiusculus Steph.	*Cerc con*	oa-ws		9	0	0	0	0	0	9	0	0	0	0	0
spp.	*Cerc spp*	rt		248	39	70	2	5	22	1	0	0	99	10	37
Megasternum boletophagum (Marsh.)	*mega bol*	rt		57	12	14	0	1	8	8	2	0	9	3	0
Cryptopleurum minutum (F.)	*Crypt min*	rf	st	43	0	3	0	0	3	0	2	0	33	2	0
Hydrobius fuscipes (L.)	*Hydr fus*	oa-w		9	2	2	0	0	0	1	2	1	1	0	0
Anacaena globulus (Payk.)	*Anac glo*	oa-w		1	1	0	0	0	0	0	0	0	0	0	0
Laccobius spp.	*Lacc spp*	oa-w		3	0	1	0	0	0	0	1	0	1	0	0
Enochrus spp.	*Enoc spp*	oa-w		2	0	1	0	0	0	0	1	0	0	0	0
Cymbiodyta marginella (F.)	*Cymb mar*	oa-w		1	0	1	0	0	0	0	0	0	0	0	0
Chaetarthria seminulum (Hbst.)	*Chae sem*	oa-w		7	7	0	0	0	0	0	0	0	0	0	0
Histeridae															
Teretrius picipes (F.)	*Tere pac*	1		2	0	2	0	0	0	0	0	0	0	0	0
Onthophilus striatus (Forst.)	*Onth str*	rt	sf	5	1	0	0	0	1	0	0	1	2	0	0
Acritus nigricornis (Hoffm.)	*Acri nig*	rt	st	97	9	36	0	2	7	10	11	2	15	5	0
Gnathoncus sp.	*Gnat spp*	rt	sf	52	0	0	0	0	7	1	15	1	3	25	0
Saprinus sp.	*Sapr spp*	rt	sf	1	0	0	0	0	0	1	0	0	0	0	0
Dendrophilus punctatus (Hbst.)	*Dend pun*	rt	sf	5	0	4	0	0	0	0	0	0	1	0	0
Dendrophilus pygmaeus (L.)	*Dend pyg*	rt		1	0	0	0	0	0	0	0	0	1	0	2
Kissister minimus (Aubé)	*Kiss min*	rt	sf	2	2	0	0	0	0	0	0	0	0	0	0
Carcinops pumilio (Er.)	*Carc pum*	rt	sf	7	0	0	0	0	0	7	0	0	0	0	0
Paromalus flavicornis (Hbst.)	*Paro flav*	rt	sf	3	0	1	0	0	0	0	2	0	0	0	0
Hister striola Sahlb.	*Hist stri*	rt	sf	95	0	1	0	0	0	0	11	0	0	83	4
Hister cadaverinus Hoffm.	*Hist cad*	rt	sf	12	0	1	2	0	2	0	0	7	0	0	0
Hister bissexstriatus F.	*Hist bis*	rt	sf	2	0	2	0	0	0	0	0	0	0	0	0
Hister spp.	*Hist spp*	rt	sf	4	1	1	0	1	1	0	0	0	0	0	0
Atholus bimaculatus (L.)	*Atho bim*	rt	sf	15	1	5	0	0	0	3	0	0	1	5	0
Atholus duodecimstriatus (schrk.)	*Atho duo*	rt	sf	6	0	2	0	0	0	0	0	0	4	0	0
Silphidae															
Necrophorus humator (Gled.)	*Necr hum*	rf		1	0	0	0	0	1	0	0	0	0	0	0
Phosphuga atrata (L.)	*Phos atr*	rf		1	1	0	0	0	0	0	0	0	0	0	0
Silpha spp.	*Silp spp*	rt		1	0	1	0	0	0	0	0	0	0	0	0
Catopidae															
Catops spp.	*Cato spp*			19	1	2	1	0	11	0	1	3	0	0	2
Choleva sp.	*Chol spp*	oa		1	0	0	0	0	0	0	0	0	1	0	0
Clamdidae															
Clambus spp.	*Clam spp*			9	0	0	1	0	2	0	5	1	0	0	0
Colonidae															
Colon spp	*Colo spp*	oa		1	0	0	0	0	0	0	0	0	1	0	0
Scydmaenidae															
Scydmaenidae Gen. & spp. indet.	Scyd spp			12	1	2	0	0	1	1	1	0	6	0	0
Orthoperidae															
Corylophus cassidoides (Marsh.)	*Cory cas*	rt		4	0	0	0	0	0	0	0	0	4	0	0
Orthoperus spp.	*Orth spp*	rt		33	2	1	0	0	2	0	21	1	6	0	0
Ptiliidae															
Ptiliidae Genus & spp. indet.	Ptil spp	rt		38	0	8	0	0	2	0	5	0	20	3	6
Acrotrichis spp.	*Acro spp*	rt		19	3	4	0	1	0	0	6	0	5	0	0
Staphylinidae															
Micropeplus staphylinoides (Marsh.)	*Micr sta*	rt		5	3	0	0	0	0	0	1	0	1	0	0
Siagonium quadricorne Kirby	*Siag qua*			5	0	0	0	0	0	0	0	0	5	0	0
Megarthrus sp.	*mega spp*		sf	14	0	0	1	0	1	1	2	0	9	0	0
Eusphalerum sp.	*Eusp spp*		sf	1	0	1	0	0	0	0	0	0	0	0	0

	Canoco codes	Ecological codes	Synanthropic codes	Total Samples analysed	Poultry PB	Poultry 1st and 2nd century	Guildhall 2nd century	Poultry 3rd century	Poultry Saxon / Norman	Guildhall Saxon /Norman	Ball wharf Saxon/ Norman	Preachers Court Medieval	St. Johns Medieval	Winchester Palace Medieval	St. Mary's Spital / Medieval
Phyllodrepa floralis (Payk.)	*Phyl flo*	rt		83	0	1	0	0	0	0	2	1	3	76	0
Omalium riparium (Thoms.)	*Omal rip*	rt	sf	8	0	1	0	0	0	0	0	2	5	0	0
Omalium rivulare (Payk.)	*Omal riv*	rt	sf	156	0	1	2	0	14	1	43	1	0	94	4
Omalium septentrionis Thoms.	*Omal sep*	rt		5	0	2	0	0	2	0	0	0	1	0	2
O. ?allardi Fairm.Bris.	*Omal all*	rt		5	0	0	0	0	0	0	0	0	0	5	0
O. caesum Grav.	*Omal cae*	rt	st	4	1	1	0	0	0	0	1	0	1	0	23
O. excavatum Steph.	*Omal exc*	rt	sf	6	0	1	0	0	0	0	1	2	2	0	0
O. spp.	*Omal spp*	rt		160	9	5	1	1	21	5	31	11	17	59	11
Xylodromus concinnus (Marsh.)	*Xylo con*	rt-h		197	19	14	6	1	22	41	71	8	13	2	17
Olophrum spp.	*Olop spp*	oa		3	0	1	0	0	0	1	1	0	0	0	1
Acidota crenata (F.)	*Acid cre*	oa		1	0	0	1	0	0	0	0	0	0	0	0
Lesteva longelytrata (Goeze)	*Lest lon*	oa-d	st	13	0	8	0	0	0	0	5	0	0	0	1
L. spp.	*Lest spp*	oa-d	sf	15	6	2	1	0	1	1	1	0	3	0	0
Coprophilus striatulus (F.)	*Copr str*	rt	st	27	0	1	0	0	3	2	6	9	6	0	0
Trogophloeus arcuatus (Steph.)	*Trog arc*	rt		2	0	1	0	0	0	0	1	0	0	0	0
Trogophloeus bilineatus (Steph.)	*Trog bil*	rt	sf	182	10	34	1	1	3	13	116	0	4	0	0
T. rivularis Motsch.	*Trog riv*			1	0	0	0	0	0	0	0	0	1	0	0
T. fuliginosus (Grav.)	*Trog ful*	u	sf	35	3	2	0	0	1	4	23	0	2	0	0
T. corticinus (Grav.)	*Trog cor*	u		39	0	7	0	1	0	0	31	0	0	0	0
T. pusillus (Grav.)	*Trog pus*	u	st	9	9	0	0	0	0	0	0	0	0	0	0
T. elongatulus Er.	*Trog elo*	u		1	0	0	0	0	0	0	0	0	0	1	0
T. spp.	*Trog spp*	u		217	10	44	0	7	15	3	124	2	12	0	0
Aploderus caelatus (Grav.)	*Aplo cae*	rt		13	0	1	0	0	0	0	2	0	9	1	0
Oxytelus sculptus Grav.	*Oxyt scu*	rt		255	58	67	7	0	14	42	34	9	24	0	9
Oxytelus rugosus (F.)	*Oxyt rug*	rt		101	7	13	1	7	12	20	24	7	10	0	1
Oxytelus scupturatus Grav.	*Oxyt sct*	rt	sf	120	1	0	1	0	12	4	14	3	85	0	13
Oxytelus nitidulus Grav.	*Oxyt nit*	rt-d		126	19	24	0	1	10	0	67	0	3	2	0
Oxytelus complanatus Er.	*Oxyt com*	rt		38	1	1	1	0	8	11	11	1	4	0	0
tetracarinatus (Block)	*Oxyt tet*	rt		16	0	0	1	0	0	7	3	2	3	0	5
Platystethus arenarius (Fourc.)	*Plat are*	rf		69	10	7	0	1	6	17	22	0	6	0	5
Platystethus cornutus (Grav.)	*Plat cor*	oa-d		286	4	14	20	0	5	185	53	0	5	0	6
Platystethus nodifrons (Man.)	*Plat nod*	oa-d		26	1	1	0	0	0	0	24	0	0	0	0
Platystethus nitens Sahlb.)	*Plat nit*	oa-ws		2	0	0	0	0	0	1	1	0	0	0	0
Stenus spp.	*Sten spp*	u		97	20	12	0	2	5	4	41	1	8	4	7
Paederus spp.	*Paed spp*	u		6	2	1	0	0	1	0	2	0	0	0	0
Astenus spp.	*Aste spp*	rt	st	5	1	0	0	0	0	0	4	0	0	0	1
Stilicus orbiculatus (Payk.)	*Stil orb*			18	3	4	0	0	2	0	6	1	1	1	0
Lithocharis ochraceus (Grav.)	*Lith och*	rt	st	23	0	0	0	0	0	0	0	0	0	23	0
Lithocharis spp.	*Lith spp*	rt	st	117	2	8	0	1	0	3	98	2	3	0	0
Lathrobium multipunctatum Grav.	*Lath mul*	rt		1	0	0	0	0	0	0	0	0	1	0	0
Lathrobium spp.	*Lath spp*	oa	st	11	2	1	0	0	2	2	3	0	0	1	2
Phacophallus parumpunctatus (Gyll.)	*Phac par*	rt	st	9	1	7	0	0	1	0	0	0	0	0	0
Leptacinus intermedius Donisth.	*Lept int*	rt	st	15	0	0	0	0	0	0	0	0	0	15	0
Leptacinus spp.	*Lept spp*	rt	st	43	4	19	1	0	2	4	10	0	3	0	0
Gauropterus fulgidus (F.)	*Gaur ful*			1	0	0	0	0	0	0	0	0	0	1	0
Gyrohypnus fracticornis (Müll.)	*Gyro fra*	rt	st	77	15	14	0	0	6	7	22	1	11	1	3
Xantholinus spp.	*Xant spp*			125	9	1	13	1	3	35	15	1	43	4	18
Neobisnus spp.	*Neob spp*	rt		124	7	28	1	11	10	29	35	3	0	0	0
Gabrius spp.	*Gabr spp*	rt		9	2	2	0	0	1	0	2	0	2	0	0
Philonthus spp.	*Phil spp*			364	7	18	6	1	6	93	54	16	40	123	36
Quedius spp.	*Qued spp*			251	13	39	0	3	40	4	108	3	13	28	6
Staphylinus sp	*Stap spp*	u		6	0	1	0	0	2	0	3	0	0	0	0
Philonthus spp.	*Pjil spp*			4	0	0	0	0	0	0	4	0	0	0	0
Tachyporus spp.	*Tacy spp*			27	3	12	0	0	0	1	3	0	6	2	1
Tachinus subterraneus (L.)	*Tach sub*			2	0	0	0	0	0	0	0	1	1	0	0
Tachinus rufipes (Geer.)	*Tach ruf*	u	st	8	3	2	0	1	2	0	0	0	0	0	0
Tachinus spp.	*Tach spp*		sf	18	4	5	0	1	4	0	3	1	0	0	2
Leucoparyphus silphoides (L.)	*leuc sil*	rt		17	0	3	0	0	0	0	2	0	10	2	2
Falagria spp.	*Fala spp*	rt		13	2	5	0	0	0	0	6	0	0	0	0
Drusilla canaliculata (F.)	*Drus can*	rt		1	0	0	0	0	1	0	0	0	0	0	0
Aleocharinidae Genus & spp. Indet.	Aleo spp			287	34	38	5	5	19	15	94	9	35	33	8
PSELAPHIDAE															
Euplectus spp.	*Eupl spp*			10	0	1	0	0	1	0	8	0	0	0	0
Trichonyx sulcicollis (Reichb.)	*Tric sul*	u		5	0	1	0	0	0	1	1	0	2	0	0
Rybaxis sp.	*Rybr spp*			3	0	0	0	0	0	1	1	0	0	1	0
Brachygluta spp.	*Brac spp*			5	1	0	0	0	0	0	4	0	0	0	0
Cantharidae															
Cantharis sp.	*Cant spp*	oa		3	0	1	0	0	0	1	1	0	0	0	0
Malthinus biguttatus (L.)	*Malt big*	oa		1	0	1	0	0	0	0	0	0	0	0	0
Malthodes marginatus (Latr.)	*Malt mar*	oa		1	1	0	0	0	0	0	0	0	0	0	0
Melyridae															
Haplocnemus nigricornis (F.)	*Hapl nig*	l		2	0	1	0	0	0	0	1	0	0	0	0
Cleridae															

	Canoco codes	Ecological codes	Synanthropic codes	Total Samples analysed	Poultry PB	Poultry 1st and 2nd century	Guildhall 2nd century	Poultry 3rd century	Poultry Saxon / Norman	Guildhall Saxon /Norman	Ball wharf Saxon/ Norman	Preachers Court Medieval	St. Johns Medieval	Winchester Palace Medieval	St. Mary's Spital / Medieval
Necrobia violacea (L.)	*Necr vio*	rt	sf	5	0	0	0	0	0	1	0	0	2	2	1
Elateridae															
Agrlotes spp.	*Agro spp*	oa-p		3	1	1	0	0	0	0	0	0	1	0	0
Melanotus rufipes (Hbst.)	*Mela ruf*	oa-p		1	1	0	0	0	0	0	0	0	0	0	0
M. spp.	*Mela spp*	oa-p		1	0	1	0	0	0	0	0	0	0	0	0
Athous spp.	*Atho spp*	oa-p		5	0	0	0	0	0	0	2	0	3	0	0
Throscidae															
Throscus obtusus Curt.	*Thro obt*	oa-ws		1	0	0	0	0	0	1	0	0	0	0	0
Helodidae															
Helodidae Gen. & spp. Indet.	Help spp	oa-w		2	0	1	0	0	0	0	1	0	0	0	0
Dryopidae															
Dryops spp.	*Dryo spp*	oa-w		2	0	1	0	0	0	0	1	0	0	0	0
Riolus spp.	*Riol spp*	oa-w		1	0	0	0	0	0	0	0	0	0	1	0
Oulimnius spp.	*Ouli spp*	oa-w		3	0	1	0	0	0	0	2	0	0	0	0
Heteroceridae															
Heterocerus spp	*Hete spp*	oa-d		1	0	1	0	0	0	0	0	0	0	0	2
Dermestidae															
Dermestes lardarius L.	*Derm lar*	rd-h	ss	2	0	1	0	0	0	0	1	0	0	0	0
Dermestes sp.	*Derm spp*	rd-h	ss	4	0	1	0	0	0	0	2	1	0	0	0
Attagenus pellio (L.)	*Atta pel*	rd-h	ss	11	0	1	0	0	0	0	2	1	4	3	0
Athrenus spp.	*Athr spp*	rd-h	ss	3	0	1	0	0	0	0	1	0	0	1	0
Byrrhidae															
Byrrhus pilula (L.)	*Byrr pil*	oa		8	7	0	0	0	1	0	0	0	0	0	0
Ostomidae															
Tenebrioides mauretanicus (L.)	*Tene mau*	rd-h	ss	2	0	1	0	0	0	0	1	0	0	0	0
Nitidulidae															
Cateretes spp.	*Cate spp*	oa		13	3	6	0	4	0	0	0	0	0	0	0
Brachypterus urticae (F.)	*Brac urti*	oa-p		2	0	0	0	0	0	0	0	0	2	0	1
Meligethes spp.	*meli spp*	oa		14	2	1	0	0	2	1	8	0	0	0	1
Omosita discoidea (F.)	*Omos dis*	rt	sf	7	0	1	1	0	3	1	0	1	0	0	1
O. colon (L.)	*Omos col*	rt	sf	30	0	3	1	0	8	9	4	1	4	0	0
Rhizophagidae															
Rhizophagus parallelocollis Gyll.	*Rhiz par*	rt	sf	3	0	1	0	0	0	0	1	0	0	1	1
Rhizophagus spp.	*Rhiz spp*	rt	sf	23	1	4	2	0	1	4	7	1	2	1	1
Cucujidae															
Monotoma angusticollis (Gyll.)	Mono ang	rt	st	4	0	0	0	0	0	0	0	0	0	4	0
Monotoma spinicollis Aubé	*Mono spi*	rt	st	9	0	1	0	0	0	0	0	0	8	0	0
Monotoma picipes Hbst.	*Mono pic*	rt	st	21	0	0	0	0	0	15	1	0	1	4	4
Monotoma brevicollis Aubé	*Mono bre*	rt	st	1	0	1	0	0	0	0	0	0	0	0	0
M. bicolor Villa.	*Mono bic*	rt	st	7	1	2	0	0	0	1	0	0	1	2	0
M. testacea Motsch.	*Mono tes*	rt	st	1	0	0	0	0	0	0	0	0	1	0	0
M. longicollis (Gyll.)	*Mono lon*	rt	st	1	0	0	0	0	0	0	0	0	1	0	0
Monotoma spp.	*Mono spp*	rt	sf	85	8	16	0	1	5	4	14	0	25	12	2
Pediacus dermestodies (F.)	*Pedi der*	rt-h	ss	2	0	1	0	0	0	0	1	0	0	0	0
Oryzaephilus surinamensis (L.)	*Oryz sur*	g	ss	276	114	99	1	13	3	0	5	14	23	4	118
Laemophloeus ferrugineus (Steph.)	*Laem fer*	g	ss	302	145	80	5	58	2	0	7	4	1	0	0
Cryptophagidae															
Cryptophagus ?dentatus (Group)	*Cryp den*	rd-h	st	16	4	2	0	0	0	0	10	0	0	0	0
C. distinguendus Sturm	*Cryp dis*	rd-h	st	23	0	1	0	0	0	0	19	0	0	3	4
C. scanicus (L.)	*Cryp sca*	rd-h	st	39	2	5	0	0	0	0	25	0	1	6	0
C. ?scutellatus Newm.	*Cryp scu*	rd-h	st	145	6	0	0	0	0	0	139	0	0	0	0
C. spp.	*Cryp spp*	rd-h	sf	214	30	55	5	4	31	16	0	19	48	6	8
Atomaria spp.	*Atom spp*	rd-h	st	219	24	32	0	4	49	6	75	1	12	16	0
Phalacridae															
Phalacrus caricis Sturm.	*Phal car*	ws		5	1	1	0	0	0	0	3	0	0	0	0
Olibrus sp.	*Olim spp*	ws		1	1	0	0	0	0	0	0	0	0	0	0
Lathridiidae															
Lathridius nodifer (Westw.)	*Lath nod*			1	0	0	0	0	0	0	0	0	1	0	0
Enicmus minutus (Group)	*Lath min*	rd-h	st	748	66	171	13	32	81	82	222	19	50	12	15
Cartodere spp.	*Cart spp*	rd	sf	13	1	1	0	0	5	0	0	0	3	3	7
Corticaria/ corticarina spp.	*Cort spp*	rt	sf	174	19	27	0	11	25	10	57	3	16	6	11
Mycetophagidae															
Mycetophagus spp.	*Myce spp*	rf		3	1	1	0	0	1	0	0	0	0	0	0
Typhaea stercorea (L.)	*Typh ster*	rd	ss	79	7	22	0	8	5	0	35	0	2	0	3

	Canoco codes	Ecological codes	Synanthropic codes	Total Samples analysed	Poultry PB	Poultry 1st and 2nd century	Guildhall 2nd century	Poultry 3rd century	Poultry Saxon / Norman	Guildhall Saxon /Norman	Ball wharf Saxon/ Norman	Preachers Court Medieval	St. Johns Medieval	Winchester Palace Medieval	St. Mary's Spital / Medieval
Colydiidae															
Aglenus brunneus (Gyll.)	*Agle bur*	rt-h	ss	230	7	49	0	7	7	54	74	3	20	9	11
Cerylon sp.	*Cery spp*	l		2	0	1	0	0	0	0	1	0	0	0	0
Endomychidae															
Mycetaea hirta (Marsh.)	*Myce hir*	rd-h	ss	164	0	0	0	0	23	9	69	30	21	12	43
Coccinellidae															
Coccidula rufa (Hbst.)	*Cocc ruf*	oa		3	0	1	0	0	0	1	1	0	0	0	0
Thea vigintiduopunctata (L.)	*Thea vig*	oa		3	0	0	0	0	0	0	3	0	0	0	0
Lyctidae															
Lyctus linearis (Steph.)	*Lyct bru*	l-h	sf	63	9	23	6	1	5	6	11	0	2	0	0
Anobiidae															
Xestobium rufovillosum (Geer)	*Xest ruf*	l	sf	46	0	1	0	0	0	0	6	2	1	36	0
Anobium punctatum (Geer)	*Anob pun*	l-h	sf	371	6	53	20	6	32	28	170	8	22	26	12
Ptilinus pectinicornis (L.)	*Ptil pect*	l	sf	12	0	1	0	0	0	0	10	0	0	1	0
Ptinidae															
*Tipnus unicolor (*Pill. Mitt.)	*Tipn uni*	rd-h	st	286	1	6	0	1	0	0	50	15	34	179	75
Ptinus fur (L.)	*Ptin fur*	rd-h	sf	198	13	22	1	2	7	1	67	7	18	60	154
Oedemeridae															
Oedemera lurida (Marsh.)	*Oede lur*	oa		1	0	0	0	0	1	0	0	0	0	0	0
Anthicidae															
Anthicus bifasciatus (Rossi)	*Anth bif*	rt	st	1	0	0	0	0	0	1	0	0	0	0	0
Anthicus formicarius (Goeze)	*Anth for*	rt	st	77	4	16	5	0	0	11	4	0	37	0	15
Anthicus floralis (L.)	*Anth flo*	rf		2	0	0	0	0	0	0	1	0	1	0	0
Anthicus antherinus (L.)	*Anth ant*	oa		4	0	1	0	0	1	0	2	0	0	0	0
A. spp.	*Anth spp*	rt		105	7	6	0	2	0	12	14	1	63	0	21
Serrophalpidae															
Hypulus quercinus (Quensel)	*Hypu que*	l		1	0	0	0	0	1	0	0	0	0	0	0
Tenebrionidae															
Blaps mucronata Latr.	*Blap muc*	rt	ss	4	0	0	0	1	0	0	3	0	0	0	0
Palorus ratzeburgi (Wissm.)	*Palo rat*	g	ss	24	12	7	0	5	0	0	0	0	0	0	0
Tribolium castaneum (Hbst.)	*Trib cas*	g	ss	1	0	0	0	1	0	0	0	0	0	0	0
Alphitobius diaperinus (Panz.)	*Alph dia*	rf	ss	15	1	13	0	1	0	0	0	0	0	0	0
Tenebrio obscurus F.	*Tene obs*	rf	ss	14	1	5	0	1	1	0	6	0	0	0	0
Tenebrio molitor (L.)	*Tene mol*	rt	ss	5	0	1	0	0	0	0	3	0	1	0	0
Enoplopus velikensis (Pill. Mitt.)	Enop vel	l		1	0	1	0	0	0	0	0	0	0	0	0
Scarabaeidae															
Trox scaber (L.)	*Trox sca*	rt	sf	85	2	6	2	1	12	16	29	4	9	4	1
Geotrupes spp.	*Geot spp*	oa-rf		5	3	0	0	0	0	0	1	0	0	1	0
Onthophagus similis (Scriba)	*Onth sim*	oa-rf		1	1	0	0	0	0	0	0	0	0	0	0
O. spp.	*Onth spp*	oa-rf		2	1	1	0	0	0	0	0	0	0	0	0
Oxyomus silvestris (Scop.)	*Oxyo sil*	rt	st	74	17	4	3	0	4	7	1	2	36	0	2
Aphodius arenarius (Ol.)	*Apho are*	oa-rf		4	0	0	0	0	0	0	0	0	4	0	0
Aphodius rufipes (L.)	*Apho ruf*	oa-rf		2	0	0	2	0	0	0	0	0	0	0	0
Aphodius luridus (F.)	*Apho lur*	oa-rf		1	0	0	0	0	0	0	0	0	1	0	0
Aphodius contaminatus (Hbst.)	*Apho con*	oa-rf		11	4	2	0	1	3	0	0	0	0	1	0
Aphodius sphacelatus (Panz.) or *A. prodromus (*Brahm*)*	*Apho S/P*	oa-rf		36	2	6	0	1	5	2	18	0	2	0	0
A. sphacelatus (Panz.)	*Apho sph*	oa-rf		2	1	1	0	0	0	0	0	0	0	0	0
A. fimetarius (L.)	*Apho fim*	oa-rf		32	4	5	0	1	0	0	11	0	10	1	0
A. ater (Geer)	*Apho are*	oa-rf		2	0	0	0	0	0	0	2	0	0	0	0
A. granarius (L.)	*Apho gra*	oa-rf		278	14	44	49	1	13	107	24	2	20	4	0
A. spp.	*Apho spp*	oa-rf		117	3	1	0	0	1	96	2	1	13	0	9
Phyllopertha horticola (L.)	*Phyl hor*	oa-p		3	3	0	0	0	0	0	0	0	0	0	0
Cerambycidae															
Grammoptera spp.	*Gram spp*	l		1	0	0	0	0	0	0	1	0	0	0	0
Gracilia minuta (F.)	*Grac min*	l		30	0	0	0	0	8	0	22	0	0	0	0
Leiopus nebulosus (L.)	*Leio neb*	l		1	0	0	0	0	0	0	1	0	0	0	0
Chyrsomelidae															
Donacia simplex F.	*Dona sim*	oa-d		1	1	0	0	0	0	0	0	0	0	0	0
P. braccata (Scop.)	*Dona bra*	oa-d		2	0	1	0	0	0	0	1	0	0	0	0
Plateumaris sericea (L.)	*Plat ser*	oa-d		16	0	1	0	0	0	0	14	0	0	1	0
Donacia/ Plateumaris spp.	*Don / plat*	oa-d		3	0	1	0	0	0	0	2	0	0	0	0
Lema cyanella (L.)	*lema cya*	oa-g		2	0	1	0	0	0	0	1	0	0	0	0
Hydrothassa marginella (L.)	*Hydr mar*	oa-d		5	0	1	0	0	0	0	4	0	0	0	0
Prasocuris phellandrii (L.)	*Pras phe*	oa-d		16	1	2	0	0	0	1	12	0	0	0	0
Gastroidea viridula (Geer)	*Gast vir*	oa-p		2	1	0	0	0	0	0	0	1	0	0	0
Phaedon tumidulus (Germar.)	*Phae tum*	oa-p		3	0	0	0	0	0	0	0	0	3	0	0
Galerucella sp.	*Gale spp*	oa		1	0	0	0	0	1	0	0	0	0	0	0
Phyllotreta spp.	*Phyl spp*	oa		67	10	12	5	3	5	14	12	2	3	1	0
Haltica spp.	*Halt spp*	oa		2	0	1	0	0	0	0	0	0	1	0	2

	Canoco codes	Ecological codes	Synanthropic codes	Total Samples analysed	Poultry PB	Poultry 1st and 2nd century	Guildhall 2nd century	Poultry 3rd century	Poultry Saxon / Norman	Guildhall Saxon /Norman	Ball wharf Saxon/ Norman	Preachers Court Medieval	St. Johns Medieval	Winchester Palace Medieval	St. Mary's Spital / Medieval
Crepidodera sp.	*Crep spp*	oa		1	0	0	0	0	1	0	0	0	0	0	0
Chalcoides sp.	*Chal spp*	oa		1	0	0	0	0	1	0	0	0	0	0	0
Podagrica fuscipes (F.)	*Poda fus*	oa-p		1	0	0	0	0	0	0	0	0	1	0	0
Chaetocnema concinna (Marsh.)	*Chae con*	oa		17	2	4	0	0	1	0	7	0	3	0	0
C. spp.	*Chae spp*	oa		19	3	2	0	0	1	2	6	0	2	3	0
Longitarsus sp.	*Long spp*	oa		4	0	1	0	0	0	0	2	0	1	0	0
Psylliodes sp.	*Psyl spp*	oa-p		4	0	1	0	0	0	0	0	0	2	1	0
Bruchidae															
Bruchus loti (Payk.)	*Bruc lot*	oa-p		13	0	0	0	0	0	0	0	0	13	0	0
Bruchus pisorum (L.)	*Bruc pis*	oa-pu		198	0	2	0	0	184	0	12	0	0	0	0
Bruchus spp.	*Bruc spp*	oa		136	5	13	1	3	1	61	16	2	33	1	0
Scolytidae															
Scolytus rugulosus (Müll.)	*Scol rug*	oa-l		13	1	4	0	0	2	5	1	0	0	0	0
Scolytus intricatus (Ratz.)	*Scol int*	oa-l		1	0	0	0	0	0	0	0	1	0	0	0
Phloeophthorus rhododactylus (Marsh.)	*Phlo rho*	oa-l		1	0	0	0	0	1	0	0	0	0	0	0
Leperisinus varius (F.)	*Lepe var*	oa-l		10	3	2	0	0	1	0	2	1	1	0	0
Hylesinus oleiperda (F.)	*Hyle ole*	oa-l		2	0	1	0	0	0	0	1	0	0	0	0
Pteleobius vittatus (F.)	*Ptel vit*	oa-l		3	2	1	0	0	0	0	0	0	0	0	0
Dryocoetes villosus (F.)	*Dryo vil*	oa-l		6	0	2	0	0	0	1	2	1	0	0	0
Dryocoetes alni georg	*dryo aln*	oa-l		4	0	0	0	0	0	1	3	0	0	0	0
Xyleborus dryographus (Ratz.)	*Xylo dry*	oa-l		2	0	1	0	0	1	0	0	0	0	0	0
Curculionidae															
Apion rufirostre (F.)	*Apio ruf*	oa-p		1	0	0	0	0	0	0	0	0	1	0	0
Apion aeneum (F.)	*Apio aen*	oa		10	0	0	0	0	0	0	1	2	7	0	0
A. radiolus (Marsh.)	*Apio rad*	oa-p		3	0	0	0	0	0	0	0	0	3	0	0
Apion difficile Hbst.	*Apio dif*	oa-p		6	0	6	0	0	0	0	0	0	0	0	0
A. ulicis Forst.	*Apio uli*	oa		2	0	1	0	0	0	0	1	0	0	0	0
Apion urticarium (Hbst.)	*Apio urt*	oa-p		2	0	0	0	0	0	1	0	0	1	0	0
A. meliloti Kirby	*Apio mel*	oa		2	0	1	0	0	0	0	1	0	0	0	0
A. subulatum Kirby	*Apio sub*			1	0	0	0	0	0	0	0	0	1	0	0
A. cerdo Gerst	*Apio cer*			1	0	0	0	0	0	0	0	0	1	0	0
A. spp.	*Apio spp*	oa-p		101	13	23	2	6	8	13	12	3	19	2	15
Otiorhynchus sulcatus (F.)	*Otio sul*	oa		0	0	0	0	0	0	0	0	0	0	0	1
Otiorhynchus ovatus (L.)	*Otio ova*	oa		1	0	0	0	0	0	0	0	0	1	0	0
Phyllobius argentatus (L.)	*Phyl arg*	oa-p		2	0	0	0	0	2	0	0	0	0	0	0
Phyllobius sp.	*Phyllob*	oa-p		2	0	0	0	1	0	0	1	0	0	0	0
Barypeithes spp.	*Bary spp*	oa		2	0	1	0	0	0	0	1	0	0	0	0
Strophosoma melanogrammum (Forst.)	*Stro mel*	oa-p		1	1	0	0	0	0	0	0	0	0	0	0
S. sp.	*Stro spp*	oa-p		1	0	1	0	0	0	0	0	0	0	0	2
Sitona lineatus (L.)	*Sito liin*	oa-p		13	2	1	0	1	0	0	4	0	3	2	0
S. suturalis Steph.	*Sito sut*	oa		2	0	1	0	0	0	0	1	0	0	0	0
S. puncticollis Steph.	*Sito pun*	oa-p		1	0	0	0	0	0	1	0	0	0	0	0
S. flavescens (Marsh.)	*Sito fla*	oa-p		11	1	1	0	0	0	9	0	0	0	0	0
S. hispidulus (F.)	*Sito his*	oa		3	0	1	0	0	0	0	1	0	1	0	0
S. humeralis Steph.	*Sito hum*	oa-p		3	1	0	0	1	0	0	0	1	0	0	0
S. spp.	*Sito spp*	oa		30	10	3	0	1	2	3	7	2	2	0	2
Pselactus spadix (Hbst.)	*Psel spa*	l		1	1	0	0	0	0	0	0	0	0	0	0
Rhyncolus spp.	*Rhyn spp*	oa-l		1	0	1	0	0	0	0	0	0	0	0	0
Bagous spp.	*Bago spp*	oa-d		5	2	1	0	0	0	0	2	0	0	0	0
Tanysphyrus lemnae (Payk.)	*Tany lem*	oa-w		3	0	1	0	0	0	0	1	1	0	0	0
Notaris acridulus (L.)	*Nota ari*	oa-d		10	4	1	0	0	1	1	3	0	0	0	0
Thyrogenes spp.	*Thyr spp*	oa-d		10	0	0	0	0	1	0	8	0	0	1	0
Curclio salicivorus Payk.	*Curc sal*	oa-l		2	0	1	0	0	0	0	1	0	0	0	0
Magdalis sp.	*Maga spp*	oa-l		1	0	1	0	0	0	0	0	0	0	0	0
Leiosoma deflexum (Panz.)	*Leio def*	oa		1	0	0	0	0	1	0	0	0	0	0	0
Hypera zoilus (Scop.)	*Hype zol*	oa-p		4	1	2	0	0	0	0	0	0	1	0	0
H. spp.	*Hype spp*	oa-p		15	1	2	1	0	0	1	6	0	3	1	0
Gronops lunatus (F.)	*Gron lun*	oa		2	0	1	0	0	0	0	1	0	0	0	0
Limnobaris pilistriata (Steph.)	*Limn pil*	oa-d		5	1	2	0	0	0	0	2	0	0	0	0
Rhyncolus chloropus (L.)	*Rhyn chl*	l		5	0	1	0	0	0	0	4	0	0	0	0
Phloeophagus lignarius (Marsh.)	*Phlo lig*	l		9	0	1	0	0	0	0	8	0	0	0	0
Sitophilus granarius (L.)	*Sito gran*	g	ss	347	82	73	2	109	5	1	12	7	50	6	307
Rhinocus pericarpius (L.)	*Rhin per*	oa-p		1	0	0	0	0	1	0	0	0	0	0	0
Rhinocus spp.	*Phin spp*	oa-p		6	1	0	0	1	2	0	2	0	0	0	0
Zacladus affinus (Payk.)	*Zacl aff*	oa-p		1	1	0	0	0	0	0	0	0	0	0	0
Ceutorhynchus contractus (Marsh.)	*Ceut con*	oa-p		8	1	0	0	0	0	2	2	3	0	0	0
C. eryisimi (F.)	*Ceut ery*	oa-p		5	0	1	1	0	0	0	2	0	1	0	0
C. ? pollinarius Forst.	*Ceut pol*	oa-p		1	0	0	0	0	0	0	0	0	1	0	0
C. spp.	*Ceut spp*	oa-p		14	1	2	0	0	1	0	4	0	6	0	3
Cidnorhinus quadrimaculatus (L.)	*Cidn qua*	oa-p		9	0	2	0	0	1	0	2	0	4	0	0
Mecinus pyraster (Hbst.)	*Meci pyr*	oa-p		3	1	2	0	0	0	0	0	0	0	0	0
Gymnetron pascuorum (Gyll.)	*Gymn pas*	Oa-p		22	0	0	0	0	0	1	3	0	18	0	0
Gymnetron villosulum Gyll.	*Gymn vil*	oa-p		2	1	1	0	0	0	0	0	0	0	0	0
G. spp.	*Gymn spp*	oa-p		46	2	6	0	4	1	0	4	1	28	0	0
Rhynchaenus sp.	*Rhyn spp*	oa-l		1	0	1	0	0	0	0	0	0	0	0	0

	Canoco codes	Ecological codes	Synanthropic codes	Total Samples analysed	Poultry PB	Poultry 1st and 2nd century	Guildhall 2nd century	Poultry 3rd century	Poultry Saxon / Norman	Guildhall Saxon /Norman	Ball wharf Saxon/ Norman	Preachers Court Medieval	St. Johns Medieval	Winchester Palace Medieval	St. Mary's Spital / Medieval
Rhynchaenus fagi (L.)	*Rhyn fag*	oa-l		2	0	1	0	0	0	0	1	0	0	0	0
MALLOPHAGA															
Damalina capris Gurlt	*Dama cap*			2	0	0	0	0	0	0	0	0	2	0	0
Damalina spp.	*Dama spp*			6	0	0	0	0	0	0	0	0	6	0	0
SIPHONAPTERA															
Pulex irritans (L.)	*Pull irr*			3	0	1	0	0	0	1	0	0	1	0	0
Ctenocephalides canis L.	*dog flea*			1	0	0	0	0	0	0	0	0	1	0	0
DIPTERA															
SUBORDER NEMATOCERA															
Family, genus & spp. indet.				167	1	5	0	0	27	134	0	0	0	0	0
SUBORDER CYCLORRHAPHA															
Family, genus & spp. indet.				215	3	37	2	0	69	0	66	0	23	15	0
Lonchopteridae															
?Lonchoptera spp.	*Lonc spp*			2	0	0	0	0	2	0	0	0	0	0	0
Syrphidae															
Eristalis ?tenax (L.)	*Eris ten*			51	0	0	0	0	9	0	0	42	0	0	0
Helomyzidae															
Heleomyza serrata (L.)	*Hele ser*			38	0	3	1	2	0	2	0	0	0	30	0
Sepsidae															
Sepsis spp.	*Seps spp*			536	7	26	104	0	4	371	15	6	3	0	0
Sphaeroceridae															
Sphaerocera curvipesLat.	*Spha cur*			125	0	0	0	0	0	3	0	8	114	0	0
Ischiolepta cf. pusilla (Fal.)	*Isch pus*			196	1	47	3	0	0	2	0	0	143	0	0
Copromyzinae Genus and spp. indet.	Copr spp			96	2	40	17	0	0	0	18	1	8	10	0
Limosininae Gen. & spp. Indet.	Limo spp			8	0	0	0	0	0	8	0	0	0	0	0
Opalimosina spp.	*Opal spp*			2	0	0	0	0	0	2	0	0	0	0	0
cf. *Telomerina flavipes* (Meigen)	Telo fla			484	8	44	13	3	151	0	21	26	40	178	0
Thoracochaeta zosterae (Hal.)	*Thor zos*			1594	0	0	2	0	502	193	581	282	0	34	0
Drosophilidae															
Drosophilia sp.	*Dros spp*			1	0	0	0	0	1	0	0	0	0	0	0
Sarcophagidae															
Sarcophaga spp.	*Sarc spp*			1	0	1	0	0	0	0	0	0	0	0	0
Calliphoridae															
Calliphora spp.	*Call spp*			10	0	3	1	0	0	4	1	0	0	1	0
Scathophagidae															
?Scathophaga sp.	Scat spp			1	1	0	0	0	0	0	0	0	0	0	0
Fanniinae															
Fannia scalaris (L.)	*Fann can*			14	0	0	0	0	0	0	12	2	0	0	0
Muscinae															
Musca domestica L.	*Musc dom*			355	27	69	6	33	0	130	55	0	0	35	0
Hydrotaea ?dentipes (F.)	*Hydr den*			21	1	12	0	0	1	0	0	0	0	7	0
Muscina stabulans (Fall.)	*Musc sta*			48	0	0	0	0	4	15	22	0	0	7	0
Stomoxys calcitrans (L.)	*Stom cal*			77	1	16	0	1	2	34	15	0	0	8	0
Hippoboscidae															
Melophagus ovinus L.	*Melo ovi*			4	0	0	0	0	1	2	1	0	0	0	0
TRICOPTERA															
Genus and spp. Indet.				8	0	1	0	0	1	1	5	0	0	0	0
HYMENOPTERA															
Formicoidea Family Genus and spp. indet.				85	15	6	0	0	28	0	36	0	0	0	0
CLADOCERA															
Daphnia Genus & spp. indet.		-		4	0	1	0	0	0	0	3	0	0	0	0

REFERENCES

Alfieri, A. 1931. Les insectes de la tombe de Toutankhamon. *Bulletin de la Société Royale Entomologique d'Égypte* 3/4, 188–189.

Allen, M.J. 1997 'Environment and Land use: The economic development of the communities who built Stonehenge (an economy to support the stones)' pp. 115–144 in B.W. Cunliffe and C.A. Renfrew (eds.) *Science and Stonehenge* (Proceedings of the British Academy 92). London: British Academy.

Allen, M.J. 2000 'High resolution mapping of Neolithic and Bronze Age chalklands and Landuse: The combination of multiple palaeoenvironmental analysis and topographic modelling' pp. 9–26 in A. S. Fairbairn (ed.) *Plants in Neolithic Britain and Beyond* (Neolithic Studies Group Seminar Papers 5) Oxford: Oxbow Books.

Allison, E. P., Kenward, H. K., Large, F. and Morgan, L. 1989. *Insect remains from the Castle Street, Carlisle site.* Prepared for Carlisle Archaeological Unit. [Environmental Archaeology Unit, York Report **89/3**]

Ashbee P., Bell, M. and Proudfoot, E. *Wilsford Shaft. Excavations 1960–62.* London: English Heritage.

Aitken, M. J. 1990. *Science-based Dating in Archaeology.* London: Longman.

Ashworth, A.C. 2004. Quaternary Coleoptera of the United States and Canada. *Developments in Quaternary Science* 1, 505–517.

Ashton, N., Lewis, S. G. and Stringer, C. 2006. Introduction: The Palaeolithic occupation of Europe. A tribute to John J. Wymer, 1928–2006. *Journal of Quaternary Science* 21, 421–424.

Association for Environmental Archaeology. 1995. *Environmental Archaeology and Archaeological evaluations: Recommendations concerning the Environmental Archaeology Component of Archaeological Evaluations in England* (Working Papers of the Association for Environmental Archaeology Number 2). York: Association for Environmental Archaeology.

Atkinson, T. C., Briffa, K. R. & Coope, G. R. 1987. Seasonal temperatures in Britain during the past 22,000 years, reconstructed using beetle remains. *Nature* 325, 587–592.

Austin, P. 2000. 'The emperor's new garden: Woodland trees, and people in the Neolithic of Britain' in A. S. Fairbairn (ed.) *Plants in Neolithic Britain and Beyond* (Neolithic Studies Group Seminar Papers 5) Oxford: Oxbow Books.

Barber, B. & Thomas C. 2002. *The London Charterhouse* (Museum of London Archaeology Service Monograph 10). London: Museum of London Archaeology Service.

Bateman, N. 1997. The London amphitheatre excavations 1987–1996. *Britannia* 28, 51–85.

Bateman, N. 1990. The discovery of Londinium's amphitheatre: Excavations at the Old Art Gallery site 1987–88 and 1990. *London Archaeologist* 6, 232–241.

Bateman, N., Cowan, C. and Wroe-Brown, R. 2008. *London's Roman Amphitheatre: Guildhall Yard, City of London* (Museum of London Archaeology Service Monograph 10). London: Museum of London Archaeology Service.

Bennett, K.D., Fossitt, J.A., Sharp, M.J. and Switsur, V.R. 1990. Holocene vegetation and environmental history at Loch Lang, South Uist, Scotland. *New Phytologist* 114, 281–298.

Bell, A. 1920. Notes on the later Tertiary Invertebrata. *Annual Report of the Yorkshire Philosophical Society* (1920), 1–21.

Bell, A. 1922. On the Pleistocene and Later Tertiary British insects. *Annual Report of the Yorkshire Philosophical Society* (1921), 41–51.

Bell, M. 2008. *Prehistoric Coastal Communities: The Mesolithic in Western Britain* (Council for British Archaeology Research Report 149). York: Council for British Archaeology.

Bell, M., Caseldine, A. and Neumann, H. 2000. *Prehistoric Intertidal Archaeology in the Welsh Severn Estuary* (Council for British Archaeology Research Report 120). London: Council for British Archaeology.

Bell, M. and Walker, M.J.C. 2005. *Late Quaternary Environmental Change* (Second edition). Harlow: Pearson.

Belshaw, R. 1989. A note on the recovery of *Thoracochaeta zosterae* (Haliday) (Diptera: Sphaeroceridae) from archaeological deposits. *Circaea* 6, 39–41.

Biddle, M. 1984. London on the Strand. *Popular Archaeology* (July), 23–27.

Binford, L. R. 1981. Behavioural archaeology and the 'Pompeii premise'. *Journal of Anthropological Research* 37, 195–208.

Blackmore, L. Bowsher, D., Cowie, R and Malcolm, G. 1998. Royal Opera House. *Current Archaeology* 14, 60–63.

Blatherwick, S. 2000. 'The archaeology of entertainment: London's Tutor and Stuart Playhouses' pp. 252–271 in I. Haynes, H. Sheldon and L. Hannigan (eds.) *London Underground: The Archaeology of a City.* Oxford: Oxbow Books.

Blair, K. G. 1924. Some coleopterous remains from the peat bed at Wolvercote, Oxfordshire. *Transactions of the Royal Entomological Society of London* 71, 558.

Blair, K. G. 1935. Beetle remains from a block of peat on the coast of East Anglia. *Proceedings of the Royal Entomological Society of London* 10, 19–20.

Bluer, R. and Brigham, T. 2006. *Roman and Later Development East of the Forum and Cornhill: Excavations at Lloyd's Register, 71 Fenchurch Street, City of London* (Museum of London Archaeology Service Monograph 30). London: Museum of London.

Bogaard, A. 2004. *Neolithic Farming in Central Europe: An Archaeobotanical Study of Crop Husbandry Practices.* London: Routledge.

Bowsher, D., Dyson, T., Holder, N. and Howell, I. 2008. *The London Guildhall: An Achaeological History of a Neighbourhood from Early Medieval to Modern Times.* (Museum of London Archaeology Service Monograph 36). London: Museum of London Archaeology Service.

Bradley, R. 1997. *The Significance of Monuments.* London: Routledge.

Bradley, R. 2000. *An Archaeology of Natural Places*. London: Routledge.

Bradley, R. 2006. *The Prehistory of Britain and Ireland.* Cambridge: Cambridge University Press.

Branch, N. P. and Green, C.P. 2004. 'The environmental history of Surrey' pp. 1-18 in J. Cotton, G. Crocker and A. Graham *Aspects of Archaeology and History in Surrey*. Guildford: Surrey Archaeological Society.

Bray, P.J., Blockley, S.P.E., Coope, G.R., Dadswell, L.F., Elias, S.A., Lowe, J.J., Pollard, A.M., 2006. Refining Mutual Climatic Range (MCR) quantitative estimates of palaeotemperature using *Ubiquity Analysis*. *Quaternary Science Reviews* 25, 1865–1876.

Bridgeland, D.R. 1994. *Quaternary of the Thames*. London: Champman and Hall.

Bridgeland, D.R., Preece, R.C., Roe, H.M., Tipping, R.M., Coope, G.R., Field, M.H. Robinson, J.E. Schreve, D.C. and Crowe, K. 2001. Middle Pleistocene interglacial deposits at Barling, Essex, U.K.: Evidence for a longer chronology for the Thames Terrace sequence. *Journal of Quaternary Science* 16, 813–840.

Brothwell, D.R. 1982. 'Linking urban man with his urban environment' pp. 126–129 in A.R. Hall and H.K. Kenward (eds.) *Environmental Archaeology in the Urban Context* (Council for British Archaeology Research Report 43). London: Council for British Archaeology.

Brothwell, D.R. 1994. 'On the possibility of urban-rural contrasts in human population palaeobiology' pp. 129–36 in A.R. Hall and H.K. Kenward, (eds.) *Urban – Rural Connexions; Perspectives from Environmental Archaeology* (Symposia of the Association from Environmental Archaeology 12). Oxford: Oxbow Books.

Brown, A.G. 2000. 'Floodplain vegetation history: Clearings as potential ritual spaces?' pp. 49–62 in A. S. Fairbairn (ed.) *Plants in Neolithic Britain and Beyond* (Neolithic Studies Group Seminar Papers 5) Oxford: Oxbow Books.

Brown, N, and Cotton, J. 2000. 'The Bronze age' pp. 81–100 in Museum of London *The Archaeology of Greater London: An assessment of Archaeological Evidence for Human Presence in the Area now Covered by Greater London.* London: Museum of London Archaeology Service.

Buckland, P. C. 1978. 'Cereal production, storage and population: A caveat' pp 43–45 in S. Limbrey & J. G. Evans (eds.) The *effect of Man on the Landscape: The Lowland Zone* (Council for British Archaeology Research Report 21), London: Council for British Archaeology.

Buckland, P.C. 1979. *Thorne Moors: A Palaeoecological Study of a Bronze Age Site: A Contribution to the History of the British Insect Fauna* (Department of Geography, University of Birmingham, Occasional Publication Number 8). Birmingham: University of Birmingham.

Buckland, P. C. 1982. The Malton burnt grain: A cautionary tale. *Yorkshire Archaeological Journal* 54, 53–61.

Buckland, P.I. and Buckland, P.C. 2006. *Bugs Colepoteral Ecology Package* (version: BUGSCEP 7.63. downloaded November 2007) www.BUGSCEP.com

Buckland, P.C. and Coope, G.R. 1991. *A Bibliography and Literature Review of Quaternary Insects.* Sheffield: J.R. Collis .

Buckland, P. C. & Perry, D. W. 1989. Ectoparasites of sheep from Storaborg, Iceland and their interpretation: Piss, parasites and people, a palaeoecological perspective. *Hikuin* 15, 37–46.

Buckland, P.C., Greig, J. R. A. & Kenward, H. K. 1974. York: An early Medieval site. *Antiquity* 48, 25–33.

Buckland, P.C., Sadler, J.P. & Smith, D. 1993. 'An insect's eye-view of the Norse farm' pp. 518–28, in C. Batey, E.J. Jesch, and C.D. Morris (eds.) *The Viking Age in Caithness, Orkney and the North Atlantic.* Edinburgh: University of Edinburgh Press.

Buckland, P.C., Sadler, J.P. & Sveinbjarnardóttir, G. 1992. 'Palaeoecological investigations at Reykholt, Western Iceland' pp. 149–168 in, C.J.Morris & D.J.Rackman (eds.) *Norse and Later Settlement and Subsistence in the North Atlantic.* Glasgow; Department of Archaeology, University of Glasgow.

Buckland, P. C., Sveinbjarnardóttir, G., Savory, D., Mcgovern, T. H., Skidmore, P. & Andreasen, C. 1983. Norsemen at Nipáitsoq, Greenland: A Palaeoecological Investigation. *Norwegian Archaeological Review* 16, 86–98.

Burch, M, and Treveil, P, with Keene, D, 2011. *The Development of Early Medieval and Later Poultry and Cheapside: Excavations at 1 Poultry and Vicinity, City of London* (Musem of London Archaeology Service Monograph Series 38). London: Museum of London.

Buxton K., Howard-Davis C., 2000 *Bremetenacum. Excavations at Roman Ribchester 1980, 1989–1990* (Lancaster Imprints Series No.9): Lancaster: Lancaster Imprints.

Carrott, J. and Kenward H.K. 2001. Species associations amongst insect remains from urban archaeological deposits and their significance in reconstructing the past human environment. *Journal of Archaeological Science* 28, 887–905.

Caseldine, A. 2000. 'The vegetation history of the Goldcliff area' pp. 208–244 in M. Bell, A. Caseldine, and H. Neumann (eds.). *Prehistoric Intertidal Archaeology in the Welsh Severn Estuary* (Council for British Archaeology Research Report 120). London: Council for British Archaeology.

Caseldine, C., Fyfe, R. and Hjelle, K. 2008. Pollen modelling, palaeoecology and archaeology: Virtualisation and/ or visualisation of the past? *Vegetation History and Archaeobotany* 17, 543–549.

Chapman, H. and Johnson, T. 1973. Excavations at Aldgate and Bush Lane House in the City of London 1972. *Transactions of the London and Middlesex Archaeology Society* 24, 71–73.

Chambers, F.M., Mighall, T.M. and Keen D.H. 1996. Early Holocene pollen and molluscan record from Enfield Lock, Middlesex, UK. *Proceedings of the Geologists Association* 107, 1–14.

Clark, J. 1989. *Saxon and Norman London* (2nd edition). London: Museum of London.

Clark, J. 2000 'Late Saxon and Norman London: Thirty years on' pp. 206–222 in I. Haynes, H. Sheldon and L. Hannigan (eds.) *London Underground: The Archaeology of a City*. Oxford: Oxbow Books.

Clark, S. & Edwards, K. 2004. Elm bark beetle in Holocene peat deposits and the northwest European elm decline. *Journal of Quaternary Science* 19, 525–528.

Coombs, C.W. & Freeman, J.A. 1956. The insect fauna of an empty granary. *Journal of Entomological Research* 46, 399–417.

Coombs, C.W. and Woodroffe, G.E. 1963. An experimental demonstration of ecological succession in an insect population breeding in stored wheat. *Journal of Animal Ecology* 32, 271–279.

Conneller, C. 'Lithic Technology and the *chaîne opératoire*' pp. 160–176 in J. Pollard (ed.) *Prehistoric Britain.* Oxford: Blackwell Publishing.

Coope, G.R. 1959. A late Pleistocene insect fauna from Chelford, Cheshire. *Proceedings of the Royal Society of London* (series B) 151, 70–86.

Coope, G. R. 1961. On the study of glacial and interglacial insect faunas. *Proceedings of the Linnean Society of London* 172, 62–65.

Coope, G. R. 1973. Tibetan species of dung beetle from Late Pleistocene deposits in England. *Nature* 245, 335–336.

Coope, G.R. 1977. Fossil Coleoptera assemblages as sensitive indicators of climatic changes during the Devensian (Last) cold stage. *Philosphical Transactions of the Royal Society of London* B 280, 313–340.

Coope, G. R. 1978. 'Constancy of insect species versus inconstancy of Quaternary environments' pp 176–187 in L. A. Mound & N. Waloff (eds.) *Diversity of Insect Faunas* (Symposia of the Royal Entomological Society of London, 9). Oxford: Blackwell.

Coope, G. R. 1981. 'Report on the Coleoptera from an eleventh-century house at Christ Church Place, Dublin' pp 51–56 in H. Bekker-Nielsen, P. Foote & O. Olsen (eds.) *Proceedings of the Eighth Viking Congress* (1977). Odense: Odense University Press.

Coope, G.R. 2001. Biostratigraphical distinction of interglacial coleopteran assemblages from southern Britain attributed to Oxygen Isotope Stages 5e and 7. *Quaternary Science Reviews* 20, 1717–1722

Coope, G.R., 2004. Several million years of stability among insect species because of, or in spite of, Ice Age climatic instability? *Philosophical Transactions of the Royal Society of London* B 359, 209–214.

Coope, G.R. 2010. Coleopteran faunas as indicators of interglacial climates in central and southern England. *Quaternary Science Reviews* 29. 1507–1514.

Coope, G.R., 2006. Insect faunas associated with Paleolithic industries from five sites of pre-Anglian age in central England. *Quaternary Science Reviews* 25, 1738–1754.

Coope, G. R. & Angus, R. B. 1975. An ecological study of a temperate interlude in the middle of the last glaciation, based on fossil Coleoptera from Isleworth, Middlesex. *Journal of Animal Ecology* 44, 365–391.

Coope, G. R. & Brophy, J. A. 1972. Late Glacial environmental changes indicated by a coleopteran succession from North Wales. *Boreas* 1, 97–142.

Coope, G. R., Gibbard, P. L., Hall, A. R., Preece, R. C., Robinson, J. E. & Sutcliffe, A. J. 1997. Climatic and environmental reconstruction based on fossil assemblages from Middle Devensian (Weichselian) deposits of the River Thames at South Kensington, Central London, UK. *Quaternary Science Reviews* 16, 1163–1195.

Coope, G. R. & Osborne, P. J. 1968. Report on the Coleopterous fauna of the Roman well at Barnsley Park, Gloucestershire. *Transactions of the Bristol and Gloucestershire Archaeological Society* 86, 84–87.

Cotton, J. 2000. 'Foragers and farmers: Towards the development of a settled landscape in London, c. 4000–1200 BC' pp. 9–34 in I. Haynes, H. Sheldon and L. Hannigan (eds.) *London Under Ground: The Archaeology of a City.* Oxford: Oxbow Books.

Cotton, J. 2004. 'Surrey's early past: A survey of recent work' pp. 19–38 in J. Cotton, G. Crocker and A. Graham (eds.) *Aspects of Archaeology and History in Surrey.* Guildford: Surrey Archaeological Society.

Cowie, R. 2000. 'Londinium to Lundenwic: Early and Middle Saxon Archaeology in the London region' pp. 175–205 in I. Haynes, H. Sheldon and L. Hannigan (eds.) *London Underground: The Archaeology of a City.* Oxford: Oxbow Books.

Cowie, R. 2003. *Urban Development in North West Roman Southwark: Excavations 1974–1990.* (Museum of London Archaeology Service Monograph Series 16). London: Museum of London.

Cowie, R. 2008. 'Descent into darkness: London in the 5th and 6th centuries' pp. 49–53 in J. Clark, J. Cotton, J. Hall, R. Sherris and H. Swain (eds.) *Londinium and Beyond: Essays on Roman London and its Hinterland for Harry Sheldon* (Council for British Archaeology Reserch Report 156). London: Council for British Archaeology.

Cowie, R. and Harding, C. 2000. 'Saxon settlement and economy from the Dark Ages to Doomsday' pp. 171–198 in Museum of London *The Archaeology of Greater London: An assessment of Archaeological Evidence for Human Presence in the Area now Covered by Greater London.* London: Museum of London.

Crockett, A. 2002. The Archaeological Landscape of Imperial College Sports Ground Part 1: Prehistoric. *London Archaeologist* 12, 295–299.

Cummings, V. and Whittle, A. 2003. Tombs with a view: Landscape, monuments and trees. *Antiquity* 77, 255–266.

Dansgraard, W., White, J.W.C. and Johnsen, J. 1989. The abrupt termination of the Younger Dryas Climate event. *Nature* 339, 532–533.

Dark, K. 1993. *Civitas to Kingdom: British Continuity 300-800.* Leicester: Leicester University Press.

Dark, K. 2000. *Britain and the End of the Roman Empire.* Stroud: Tempus.

Darvill, T. 1987. *Prehistoric Britain.* London: Batsford.

Darvill, T. 2006. *Stonehenge: The Biography of a Landscape.* Stround: Tempus.

Davis, A. 1997. 'The plant remains' pp 234–244 in C. Thomas, B., Sloane and C. Philpotts (eds.), *Excavations at the Priory and Hospital of St. Mary Spital, London* (Museum of London Archaeology Service Monograph 1). London: Museum of London.

Davis, A. 2003. 'The plant remains' pp. 289–301 in G. Malcolm, D. Bowsher and R. Cowie, 2003. *Middle Saxon London: Excavations at the Royal Opera House 1989-99* (Museum of London Archaeology Service Monograph 15). London: Museum of London.

Davis, A. 2004. 'Plant remains' pp. 367–382 in B. Sloane and G. Malcolm (eds.) *Excavations at the Priory of the Order of the Hospital of St John of Jerusalem, Clerkenwell, London* (Museum of London Archaeology Service Monograph 20). London: Museum of London.

Davis, A. 2011. 'Plant remains' pp. 318–319 in M. Burch and P. Treveil, with D. Keene (eds.) *The Development of Early Medieval and Later Poultry and Cheapside: Excavations at 1 Poultry and Vicinity, City of London* (Museum of London Archaeology Services Monograph Series 38). London: Museum of London.

Davis, A. and de Moulins, D. 1988. 'The plant remains' pp. 139–147 R. Cowie and R. L. Whytehead (eds.) Two Middle Saxon occupation sites: Excavations at Jubilee Hall and 21–22 Maiden Lane, WC1. *Transactions of the London and Middlesex Archaeology Society* 39, 47–163.

Davies, P., Robb, J.G. and Ladbrook, D. 2005. Woodland clearance in the Mesolithic: The social aspects. *Antiquity* 79, 280–288.

de Moulins, D. 1990. 'Environmental analysis' pp. 85–115 in C. Maloney *The Upper Walbrook Valley in the Roman Period* (Council for British Archaeology Research Report 69). London: Council for British Archaeology.

de Moulins, D. and Davis, A. 1989. 'The plant remains' pp 134–48 in R. L. Whytehead, R. Cowie, and L. Blackmore (eds.) Excavations at the Peabody site, Chandos Place, and the National Gallery. *Transactions of the London and Middlesex Archaeology Society* 40, 35–176.

de Rouffignac, C. 1997. 'The parasite remains' p. 247 in C. Thomas, B., Sloane and C. Philpotts (eds.), *Excavations at the Priory and Hospital of St. Mary Spital, London* (Museum of London Archaeology Service Monograph 1). London: Museum of London.

Dendy, A. and Elkington, H.D. 1920. *Report on the Effect of Air-tight Storage upon Grain Insects* (Royal Society Grain Pests Committee Report No. 6). London: Royal Society.

Devoy, R.J.N. 1979. Flandrian sea-level changes and vegetational history of the lower Thames Estuary. *Philosophical Transactions of the Society of London* B 285, 355–410.

Devoy, R.J.N. 1980. 'Post Glacial environmental change and man in the Thames estuary: a synopsis' pp 134–148 in F.H. Thompson (ed.) *Archaeology and Coastal Change*. London: Society of Antiquities.

Dobney, K., Hall, A., Kenward, H.K. and Milles, A. 1992. A working classification of sample types for environmental archaeology. *Circaea* 9, 24–26.

Douglas, M. 1966. *Purity and Danger: An Analysis of Concepts of Pollution and Taboo.* London: Routledge.

Drummond-Murray, J., Saxby, D. & Watson, B. 1997. Recent archaeological work in Bermondsey district of Southwark. *London Archaeologist* 10, 251–257.

Drummond-Murray, J., Thompson, P. and Cowan, C. 2002. *Settlement in Roman Southwark: Archaeological Excavations (1991–8) for the London Underground Limited Jubilee Line Extension Project* (Museum of London Archaeology Service Monograph Series 12). London: Museum of London.

Dunwoodie, L. 2004. *Pre-Boudican and Later Activity on the Site of the Forum: Excavations at 168 Fenchruch Street, City of London.* (Museum of London Archaeology Service Monograph Series 13). London: Museum of London.

Elias, S. A.1994. *Quaternary Insects and Their Environments.* Washington: Smithsonian Institution Press.

Elias, S.A. 2006. Quaternary beetle research: The state of the art. *Quaternary Science Reviews* 25, 1731–1737.

Elias, S.A. 2010: Advances in Quaternary Entomology. (Developments in Quaternary Sciences 12). Amsterdam: Elsevier.

Ellias, S A., Webster, L. and Amer, M. 2009. A beetle's eye view of London from the Mesolithic to Late Bronze Age. *Geological Journal* 44, 537–567.

Edmonds, M. 1995. *Stone Tools and Society: Working stone in Neolithic and Bronze Age Britain.* London: Batsford.

English Heritage 2002. *Environmental Archaeology: A Guide to Theory and Practice of Methods, from Sampling and Recovery to Post-excavation* (Centre for Archaeology Guidelines 1). London: English Heritage.

Esmonde-Cleary, A. S. 1989. *The Ending of Roman Britain.* London: Batsford.

Esmonde Cleary, A. S. 2000. 'Putting the dead in their place: Burial location in Roman Britain' pp. 127–142 in J. Pearce, M. Millet and N. Struck (eds.) *Burial, Society and Context in the Roman World.* Oxford: Oxbow Books.

Evans, C., Pollard, J. and Knight, M. 1999. Life in woods: Tree throws, 'settlement' and forest cognition. *Oxford Journal of Archaeology* 18, 241–254.

Evans, J. 2003. *Environmental Archaeology and the Social Order.* London: Routledge.

Ewing, H. E. 1924. Lice from human mummies. *Science* 60, 389–390.

Exon, S. Gaffney, V., Woodward A.Yorston, R. 2000. *Stonehenge Landscapes: Journeys through Real and Imagined Landscapes.* Oxford: Archaeopress.

Fairbairn, A.S. (ed.) 2000. *Plants in Neolithic Britain and Beyond* (Neolithic Studies Group Seminar Papers 5) Oxford: Oxbow Books.

Fleming, A. 1999. Phenomenology and the megaliths of Wales: A dreaming too far? *Oxford Journal of Archaeology* 18, 119–125.

Franks, J.W. 1960. Interglacial deposits at Trafalgar Square, London. *New Phytologist* 59, 145–52.

Franks, J.W., Sutcliffe, A.J. Kerney, M.P. and Coope, G.R. 1958. Haunt of the elephant and Rhinoceros: The Trafalgar Square of 100,000 years ago – new discoveries. *Illustrated London News* (14th of June), 1011–13.

Freeman, P. 1980. *Common Insects Pests of Stored Products.* London: British Museum (Natural History).

Gaffney, V.G., Fitch, S. and Smith, D.N. 2009. *Europe's Lost World: The Rediscovery of Doggerland.* Oxford: Archaeopress.

Gauch, H.G. 1982. *Mutivariate Analysis in Community Ecology.* Cambridge: Cambridge University Press.

Giorgi, J. 2002. 'Plant remains' pp. 105–111 in B. Barber, & C. Thomas (eds.) 2002. *The London Charterhouse.* (Museum of London Archaeology Service Monograph 10). London: Museum of London.

Giorgi, J. 2006. 'The plant remains' pp. 118–130 in Seeley, D., Phillpotts, C. and Samuel, M. (eds) *Winchester Palace: Excavations at the Southwark Residence of the Bishops of Winchester* (Museum of London Archaeology Service Monograph 31). London: Museum of London.

Giorgi, J. 2007 'The plant remains' pp.475–477 in D. Bowsher, T. Dyson, N. Holder and I. Howell, 2008. *The London Guildhall: An Achaeological History of a Neighbourhood from Early Medivceal to Modern Times.* (Museum of London Archaeology Service Monograph 36). London: Museum of London.

Gibbard, P. L., Coope, G. R., Hall, A. R., Preece, R. C. & Robinson, J. E. 1982. Middle Devensian deposits beneath the 'Upper Floodplain' terrace of the River Thames at Kempton Park, Sunbury, England. *Proceedings of the Geologists' Association* 93, 275–289.

Girling, M. A. 1977. Fossil insect assemblages from Rowland's track. *Somerset Levels Papers* 3, 51–60.

Girling, M. A. 1979a. Fossil insects from the Sweet track. *Somerset Levels Papers* 5, 84–93.

Girling, M. A. 1979b. 'The insects' pp 170–172; 414 & 466 in J. Bird, A.H.,Graham, H. Sheldon and P. Townend (eds.) *Southwark Excavations 1972-1974* (London & Middlesex Archaeological Society and Surrey Archaeological Society Joint Publication 1) London: London & Middlesex Archaeological Society/ Surrey Archaeological Society.

Girling, M. A. 1979c. The fossil insect assemblages from the Meare Lake Village. *Somerset Levels Papers* 5, 25–32.

Girling, M. A. 1980. The fossil insect assemblage from the Baker Site. *Somerset Levels Papers* 6, 36-42.

Girling, M.A.1982. 'Fossil insect faunas from forest sites' pp. 129–146 in M. Bell, & S. Limbrey (eds.) *Archaeological Aspects of Woodland Ecology* (British Archaeological Reports, International Series 146), Oxford: British Archaeological Reports.

Girling, M. A. 1984. Eighteenth century records of human lice (Pthiraptera, Anoplura) and fleas (Siphonaptera, Pulicidae) in the City of London. *Entomologist's Monthly Magazine* 120, 207–210.

Girling, M. A.1985. An 'old forest' beetle fauna from a Neolithic and Bronze Age peat deposit at Stileway. *Somerset Levels Papers* 11, 80–5.

Girling, M. A.1989a. 'Mesolithic and later landscapes interpreted from the insect assemblages of West Heath Spa Hampstead' pp. 72–89 in D. Collins & D. Lorimer (eds.) *Excavations at the Mesolithic Site on West Heath, Hampstead 1976-1981* (British Archaeological Reports 217). Oxford: British Archaeological Reports.

Girling, M. A. 1989b. 'The Insect Fauna of the Roman Well at the Cattlemarket' pp. 234–241 in A. Down (ed.) *Chichester Excavations* (Volume 6). Chichester: Philimore.

Girling, M. A. and Greig, J. R. A. 1977. Palaeoecological investigations of a site at Hampstead Heath, London. *Nature* 268, 45–47.

Girling, M. A. and Greig, J. R. A. 1985. A First Fossil Record for *Scolytus scolytus* (F.) (Elm Bark Beetle): Its Occurrence in Elm Decline deposits from London and the implications for Neolithic elm disease. *Journal of Archaeological Science* 12, 347–352.

Godwin, H. 1975. *The History of the British Fauna* (2nd Edition). Cambridge: Cambridge University Press.

Gray, L. and Giorgi, J. 2008. 'Plant remains' pp. 212–213 in N. Bateman, C. Cowan and R. Wroe-Brown, 2008. *London's Roman Amphitheatre: Guildhall Yard, City of London* (Museum of London Archaeology Service Monograph 10). London: Museum of London Archaeology Service.

Green, C.P., Branch, N.P., Coope, G.R., Field, M.H., Keen, D.H., Wells, H.M., Schwnninger, J-L. Preece, R.C., Schreve, D.C., Canti, M.G. and Gleed-Owen, C.P. 2004. Marine Isotope Stage 9 environments of fluvial deposits at Hackney, North London, UK. *Quaternary Science Reviews* 25, 86–113.

Greenwood, M. and Smith, D.N. 2005. 'A survey of Coleoptera from sedimentary deposits from the Trent Valley' pp 53–67 in D.N. Smith, M.B. Brickley and W. Smith (eds.) *Fertile Ground: Papers in Honour of Professor Susan Limbrey* (Association for Environmental Archaeology Symposia Symposia No. 22). Oxford: Oxbow Books.

Greig, J. 1982. 'Past and present limewoods of Europe' pp 25–55 in M. Bell, M. and Limbrey, S., (eds.), *Archaeological Aspects of Woodland ecology* (British Archaeological Reports International Series 146). Oxford: British Archaeological Reports.

Greig, J. 1989. 'From lime forest to heathland – five thousand years of change at West Heath Spa, Hampstead as shown by the plant remains' pp. 89–99 in D. Collins & D. Lorimer (eds.) *Excavations at the Mesolithic Site on West Heath, Hampstead 1976-1981* (British Archaeological Reports 217). Oxford: British Archaeological Reports.

Hall, A. R. & Kenward, H. K. 1980. An Interpretation of Biological Remains from Highgate, Beverley. *Journal of Archaeological Science* 7, 33–51.

Hall A.R. and Kenward H.K. 1990. *Environmental Evidence from the Colonia* (The Archaeology of York 14/6). London: Council for British Archaeology.

Hall, A. and Kenward, H. 2003. 'Can we identify biological indicator groups for craft, industry and other activities?' pp. 114–30 in P. Murphy, and P. E. J. Wiltshire (eds.) *The Environmental Archaeology of Industry* (Symposia of the Association for Environmental Archaeology 20). Oxford: Oxbow Books.

Hall, A.R.and Kenward, H.K. 2011. 'Plant and invertebrate indicators of leather production: from fresh skin to leather offcuts' pp. 9–32 in R. Thomas and Q. Mould (eds.) *Leather Tannaries: The Archaeological Evidence*. London: Archaeotype Publications.

Hall, A. R., Kenward, H. K. & Williams, D. 1980. *Environmental Evidence from Roman Deposits in Skeldergate* (Archaeology of York 14/3). London: Council for British Archaeology.

Hall, A. R., Kenward, H. K., Williams, D. & Grieg, J. R. A. 1983. *Environment and Living Conditions at Two Anglo-Scandinavian Sites* (Archaeology of York 14/4). London: Council for British Archaeology.

Hall, J. and Merrifield, R. 1986. *Roman London*. London: HMSO.

Halstead, P. and Tierney, J. 1998. Leafy hay: An enthnoarchaeological study in NW Greece. *Environmental Archaeology* 1, 71–80.

Harding, V. 2000. 'Death in the City: Mortuary Archaeology to 1800' pp. 272–283 in I. Haynes, H. Sheldon and L. Hannigan (eds.) *London Underground: The Archaeology of a City*. Oxford: Oxbow Books.

Hassall, M. 2000. 'London: The Roman city' pp 52–61 in I. Haynes, H. Sheldon and L. Hannigan (eds.) *London Underground: The Archaeology of a City*. Oxford: Oxbow Books.

Haynes, I., Sheldon, H. and Hannigan L. 2000. *London Underground: The Archaeology of a City*. Oxford: Oxbow Books.

Helbaek, H. 1952. Early crops in Southern England. *Proceedings of the Prehistoric Society* 18, 194–233.

Hellqvist, M & Lemdahl, G. 1996. Insect assemblages and local environment in the Mediaeval town of Uppsala, Sweden. *Journal of Archaeological Science* 23, 873–881.

Hill, J, and Rowsome, P. 2012. *Roman London and the Walbrook Stream Crossing: Excavations at 1 Poultry and Vicinity, City of London* (Museum of London Arcaheology Service Monograph Series 37). London: Museum of London.

Hill, J.D. 1995. *Ritual and Rubbish in the Iron Age of Wessex* (British Archaeological Reports, British Series 242): Oxford: Tempus Reparatum.

Hodges, R. 1982. *Dark Age Economics: The Origins of Towns and Trade AD 600-1000*. London: Duckworth.

Holladay, A.J. and Poole, J.C.F. 1979 Thucydides and the plague of Athens. *Classical Quarterly* 29, 282–300.

Horsman, V., Milne, C. and Milne, G. 1988. *Aspects of Saxo-Norman London I: Building and Street Development near Bilingsgate and Cheapside* (London and Middlesex Archaeology Society Special Paper 11). London: London and Middlesex Archaeology Society.

Howard, A.J. and Macklin, M.G. 1999. A generic geomorphological approach to archaeological interpretation and prospection in British river valleys: A guide for archaeologists investigation Holocene landscapes. *Antiquity* 73, 527–41.

Howe, E. 2002. *Roman Defences and Medieval Industry: Excavations at Baltic House, City of London.* (Museum of London Archaeology Service Monograph Series 7). London: Museum of London.

Howe, E. & Lakin, D. 2004. *Roman and Medieval Cripplegate, City of London: Archaeological Excavations 1992–8.* (Museum of London Archaeology Service Monograph Series 21). London: Museum of London.

Huntley, B. 1993. 'Rapid early-Holocene migration and high abundance of hazel (*Corylus acellana* L): Alternative hypotheses' pp 205–215 in F. M. Chambers (ed.) *Climatic Change and Human Impact in the Landscape*. London: Chapman and Hall.

Hunter, F.A., Tulloch, B.M. & Lamborne, M.G. 1973. Insects and mites of maltings in the East Midlands of England. *Journal of Stored Product Research* 9, 119–141.

Hurschmann, Rolf (Hamburg). 2009. '*Palimpsest*': *Brill's New Pauly*. Antiquity volumes edited by: Hubert Cancik and Helmuth Schneider . Brill, 2009. Brill Online. University of Birmingham. Consulted 12 June 2009 http://www.brillonline.nl/subscriber/entry?entry=bnp_e904490 .

Hyman, P and Parsons M.S. 1992. *A Review of the Scarce and Threatened Coleoptera of Great Britain* (U.K. Nature Conservation Volume 3/1). Peterborough: UK Joint Nature Conservation Committee.

Innes, J.B. and Simmons I.G. 1988. Disturbance and diversity: Floristic changes with pre-elm decline woodland recession in North East Yorkshire pp. 7-20 in M. Jones (ed.) *Archaeology and the Flora of the British Isles* (Oxford University Committee for Archaeology Monograph Number 14). Oxford: Oxford University Committee for Archaeology.

Iversen, J. 1941. Land occupation in Denmark's Stone Age. Dasmarks Geologiske Undersøgelse, Raekke 2/66, 7–69.

Iversen, J. 1960. Problems of the Early Post-Glacial forest development in Denmark. *Damarks Geologiske Undersøgelse, IV (Raekke*, 4). 1–32.

Jones, A.E. 1996. *25-26 Long Causeway Peterborough, Cambridgeshire: Archaeological investigations 1995–5. A Post Excavation Assessment and Research Design.* Birmingham: Birmingham University Field Archaeology Unit (Report 317.02).

Jones, A.K.G. 1983. 'Human parasite remains: Prospects for a quantitative approach' pp 79–81 in A.R. Hall and H.K. Kenward (eds.) *Environmental Archaeology in the Urban Context* (Council for British Archaeology Research Report 43). London: Council for British Archaeology.

Jones, A.K.G. 1985. 'Trichurid ova in archaeological deposits: Their value as indicators for ancient faeces' pp. 105-119 in N.R.J. Fieller, D.D. Gilberson and N. Ralph (eds.) *Palaeobiological Investigations, Research Design, Methods and Data Analysis (*Symposia of the Association for Environmental Archaeology 12). Oxford: Oxbow Books.

Jones, G. 1991. 'Numerical analysis' pp. 63–80 in W. van Zeist, K. Wasylikowa and K.-E. Behre (eds.) *Progress in Old World Palaeoethnobotany*. Rotterdam: Swets & Zeitlinger.

Jones, G. 2000. 'Evaluating the importance of cultivation and collecting in Neolithic Britain' pp. 79–84 in A. S. Fairbairn (ed.) *Plants in Neolithic Britain and Beyond* (Neolithic Studies Group Seminar Papers 5). Oxford: Oxbow Books.

Jones, G. & Rowley-Conwy, P. 2007. 'On the importance of cereal cultivation in the British Neolithic' pp. 391–419 In S. Colledge, & J. Conolly (eds.) *The Origins and Spread of Domestic Plants in Southwest Asia and Europe.* Walnut Creek, California: Left Coast Press.

Jones, G; Straker, V and Davis, A. 1991. 'Early Medieval plant use and ecology' pp. 347–88, in A. G. Vince (ed.) *Aspects of Saxon and Norman London 2: Finds and Environmental Evidence.* (London and Middlesex Archaeological Society Special Paper 12). London: London and Middlesex Archaeological Society.

Keen, D.H., Coope, G.R., Jones, R.L. Grithiths, H.I., Lewis, S.G., Bowen, D.Q. 1997. Middle Pleistocene deposits at Frog Hall Pit, Stretton on Dunsmore, Warwickshire, and their implications of the age of the type Wolstonian. *Journal of Quaternary Science* 12, 182–208.

Keene, D.J. 1982. 'Rubbish in Medieval Towns' pp. 26–30 in A.R. Hall and H.K. Kenward (eds.) *Environmental Archaeology in the Urban Context* (Council for British Archaeology Research Report 43). London: Council for British Archaeology.

Kenward, H.K. 1975a. The biological and archaeological implications of the beetle *Aglenus brunneus* (Gyllenhal) in ancient faunas. *Journal of Archaeological Science* 2, 63–69.

Kenward H.K. 1975b. Pitfalls in the environmental interpretation of insect death assemblages. *Journal of Archaeological Science* 2, 85–94.

Kenward, H.K. 1976. Further archaeological records of *Aglenus brunneus* (Gyll.) in Britain and Ireland, including confirmation of its presence in the Roman period. *Journal of Archaeological Science* 3, 275–277.

Kenward H.K. 1978. *The Analysis of Archaeological Insect Assemblages: A New Approach.* (Archaeology of York, 19/1). London: Council for British Archaeology for York Archaeological Trust.

Kenward, H.K. 1982. 'Insect communities and death assemblages, past and present' pp. 71–78 in A. R. Hall & H. K. Kenward (eds.) *Environmental Archaeology in the Urban Context* (Council for British Archaeology Research Report 43). London: Council for British Archaeology.

Kenward, H.K. 1992. Rapid recording of archaeological insect remains – a reconsideration. *Circaea* 9, 81–88.

Kenward H.K. 1997. Synanthropic Insects and the size, remoteness and longevity of archaeological occupation sites: applying concepts from biogeography to past "islands" of human occupation. *Quaternary Proceedings* 5, 135–152.

Kenward, H.K. 2004. Do insect remains from historic-period archaeological occupation sites track climate change in Northern England? *Environmental Archaeology* 9, 47–59.

Kenward, H.K. 2006. The visibility of past trees and woodland: Testing the value of insect remains. *Journal of Archeological Science* 33, 1368–1380.

Kenward, H. K., Allison, E. P., Dainton, M., Kemenes, I. K. and Carrott, J. B. 1992. *Evidence from Insect Remains and Parasite Eggs from Old Grapes Lane A, The Lanes, Carlisle: Technical Report* (Ancient Monuments Laboratory Report 78/92). London: English Heritage.

Kenward H.K. and Allison E. P. 1994. 'A preliminary view of the insect assemblages from the Early Christian Rath Site at Deer Park Farms, Northern Ireland' pp. 89–103 in J. Rackham (ed.) *Environment and Economy in Anglo-Saxon England* (Council for British Archaeology Research Report 89). London: Council for British Archaeology.

Kenward, H.K. and Allison, E.P. 1995. 'Rural origins of the urban insect fauna' pp. 55–78 in A.R. Hall, & H.K. Kenward (eds.) *Urban-Rural Connexions: Perspectives from Environmental Archaeology* (Symposia of the Association for Environmental Archaeology 12). Oxford: Oxbow Books.

Kenward, H.K., Allison, E. P., Dainton, M. Kemenés, I. K. and Carrott, J. B. 2000. 'The insect and parasite remains' pp. 81–83 in McCarthy, M. R. (ed.), *Roman and Medieval Carlisle: The Southern Lanes* (Department of Archaeological Sciences, University of Bradford, Research Report 1). Carlisle: Carlisle Archaeology Limited.

Kenward, H.K. and Carrott, J. 2006. Insect species associations characterise past occupation sites. *Journal of Archaeological Science* 33, 1452–1473.

Kenward, H.K., Dainton, M., Kemenes, I. K. and Carrott, J. B. 1992a. *Evidence from Insect Remains and Parasite Eggs from the Old Grapes Lane B site, The Lanes, Carlisle: Technical Report.* (Ancient Monuments Laboratory Report 76/92). London: English Heritage.

Kenward, H.K., Dainton, M., Kemenes, I. K. and Carrott, J. B. 1992b. Evidence from Insect Remains and Parasite Eggs from the Lewthwaites Lane A site, The Lanes, Carlisle: Technical Report. (Ancient Monuments Laboratory Report 77/92). London: English Heritage.

Kenward, H.K., Engleman, C. Robertson, A. and Large, F. 1985. Rapid scanning of urban archaeological deposits for insect remains. *Circaea* 3, 163–172.

Kenward H.K. and Hall A.R. 1995. *Biological Evidence from Anglo-Scandinavian Deposits at 16-22 Coppergate* (The Archaeology of York 14/7). London: Council for British Archaeology.

Kenward, H.K. and Hall, A.R. 1997. Enhancing bio-archaeological interpretation using indicator groups: Stable manure as a paradigm. *Journal of Archaeological Science* 24, 663–673.

Kenward, H.K. and Hall, A. R. 2006. Easily decayed organic remains in urban archaeological deposits: Value, threats, research directions and conservation, pp. 183–198 in Brinkkemper, O., Deeben, J., van Doesburg, J., Hallewas, D. Theunissen, E. M. and Verlinde, A. D. (eds.), *Vakken in vlakken. Archeologische kennis in lagen.* (Nederlandse Archeologische Rapporten 32). Amersfoort: ROB.

Kenward, H. K., Hall, A. R. and Jones, A. K. G. 1980. A tested set of techniques for the extraction of plant and animal macrofossils from waterlogged archaeological deposits. *Science and Archaeology* 22, 3–15.

Kenward, H. K., Hall, A. R. & Jones, A. K. G. 1986. *Environmental Evidence from a Roman Well and Anglian Pits in the Legionary Fortress* (Archaeology of York 14/2). London: Council for British Archaeology.

Kenward, H.K., Hall, A.R. and McComish, J.M. 2004. Archaeological implications of plant and invertebrate remains from fills of a massive Post-Medieval cut at Low Fishergate, Doncaster, UK. *Environmental Archaeology* 9, 61–74

Kenward, H.K. and Large, F. 1998. Insects in urban waste pits in Viking York: Another kind of seasonality. *Environmental Archaeology* 3, 35–54.

Kenward, H. K. and Williams, D. 1979. *Biological Evidence from the Roman Warehouses in Coney Street* (Archaeology of York 14/2). London: Council for British Archaeology.

Kerney, M.P., Gibbard, P.L., Hall, A.R. and Robinson, J.E. 1982. Middle Devensian river deposits beneath the 'upper floodplain' terrace of the river Thames at Isleworth, West London. *Proceedings of the Geologists Association* 93, 385–93.

Knight, D. and Howard. A.J. 2005. *Trent Valley Landscapes.* Kings Lynn: Heritage, Marketing and Publications Ltd.

Knowles, D. 1969. 'The London Charterhouse' pp.159–69 in J.S. Cockburn, H.P.F. King and K.G.T. McDonnell (eds.) *The Victoria History of Middlesex* (Volume 1). London: Dawsons.

Kolbe. H. 1894. Über fossile Reste von Coleopteren aus einen alten Torflager (Schmierkohle) bei Gr. Räschen in der Niederlausitz. *Sitzberichte naturforschender Freunde zu Berlin* (1894), 236–238.

Kolbe, H. 1925. Vergleichender Blick auf die rezente und fossile Insektenwelt Mitteleuropas, und eine Erinnerung an meine abhandlung uber "Problematische Fossilien aus dem Culm". *Deutsche entomologische Zeitschrift* 125, 147–162.

Langdon, P.G., Barber, K.E., & Lomas-Clarke S.H. 2004. Reconstructing climate and environmental change in Northern England through chironomid and pollen analyses: Evidence from Talkin Tarn, Cumbria, *Journal of Paleolimnology* 32, 197–213.

Leroi-Gourhan, A. 1993 [1964] *Gesture and Speech.* Cambridge, MA: MIT Press.

Letts, J.B. 1999. *Smoke Blackened Thatch.* London/ Reading: English Heritage/ The University of Reading.

Lewis, J. 2000a. 'The Upper Palaeolithic and Mesolithic periods' pp. 45–61 in Museum of London *The Archaeology of Greater London: An Assessment of Archaeological Evidence for Human Presence in the Area now Covered by Greater London.* London: Museum of London.

Lewis, J. 2000b. 'The Neolithic period' pp. 83–80 in Museum of London *The Archaeology of Greater London: An Assessment of Archaeological Evidence for Human Presence in the Area now Covered by Greater London.* London: Museum of London.

Lewis, J.S.C., Wiltshire, P.E.J. and Macphail, R.I. 1992. 'A late Devensian/ early Flandrian site at Three Ways Wharf, Uxbridge: Environmental implications' pp 235–247 in S. P. Needham and M.G. Macklin (eds.) *Alluvial Archaeology in Britain* (Oxbow Monograph 27). Oxford: Oxbow Books.

Lewis, J., Brown, F., Batt, A., Cooke, N., Barrett, J. Every, R., Mepham, L., Brown, K., Cramp, K. Lawson, A.J., Roe, F., Allen, S., Petts, D., McKinley, J., Carruthers, W., Challinor, D., Wiltshire, P., Robinson, M., Lewis, H. and Bates, M. 2006. *Landscape Evolution in the Middle Thames Valley: Heathrow Terminal 5 Excavations: Volume 1, Perry Oaks.* (Framework Archaeology Monograph No.1). Oxford and Salisbury: Framework Archaeology.

Lindroth, C.H. 1974. *Coleoptera: Carabidae* (Handbooks for the Identification of British Insects 4/2). London: Royal Entomological Society of London.

Lomnicki, A. M. 1894. Fauna pleistocenica insectorum boryslaviensium. *Wydawnictwa Muzeum imienia Dzieduszyckich* 4, 1–116.

Lowe, J.J. and Walker, M.J.C. 1997a. *Reconstructing Quaternary Environments* (2nd edition). London: Longman.

Lowe, J.J. and Walker, M.J.C. 1997b. Temperature variations in NW Europe during the last glacial-interglacial transition based upon the analysis of coleopteran assemblages. *Quaternary Proceedings* 5, 165–176.

Lowe, J. J., Coope, G. R., Sheldrick, C., Harkness, D. D. & Walkwer, M. J. C. 1995. Direct comparison of UK temperatures and Greenland snow accumulation rates, 15,000–12,000 yr ago. *Journal of Quaternary Science* 10, 175–180.

Lucht, W.H. 1987. *Die Käfer Mitteleuropas* (Katalog). Krefeld: Goecke and Evers.

MacArthur, R. H. and Wilson, E. O. 1967. *The Theory of Island Biogeography.* Princeton: Princeton University Press.

MacArthur, W. 1958. The plague of Athens. *Bulletin of Historical Medicine* 32, 242–246.

Macphail, R.I., Galinié, H. and Verhaeghe, F. 2003. A future for dark earth? *Antiquity* 77, 349–358.

Malcolm, G., Bowsher, D. and Cowie, R. 2003. *Middle Saxon London: Excavations at the Royal Opera House 1989-99* (Museum of London Archaeology Service Monograph 15). London: Museum of London.

Maloney, C. 1990. *The Upper Walbrook Valley in the Roman Period* (Council for British Archaeology Research Report 69). London: Council for British Archaeology.

Manchester, K. 1992. 'The palaeopathology of urban infections' pp 8–14 in S. Bassett (ed.) *Death in Towns: Urban Responses to the Dying and the Dead 100–1600.* Leicester: Leicester University Press.

Marsden, P. 1980. *Roman London.* London: Thames and Hudson.

Marsden, P. 1987. *The Roman Forum site in London: Discoveries before 1985.* London: HMSO.

Marsden, P. and West, B. 1992. Population change in Roman London. *Britannnia* 23, 133–40.

Mattingly, D. 2006. *An Imperial Possession: Britain in the Roman Empire.* London: Penguin.

Meddens, F.M. 1996. Sites from the Thames estuary wetlands, England, and their Bronze age use. *Antiquity* 70, 325–334.

Meddens, F. and Sidell, E.J. 1995. Bronze Age trackways in east London. *Current Archaeology* 12, 412–416.

Mellars, P.A. 1976. Fire ecology, animal populations and man: A study of some ecological relationships in prehistory. *Proceedings of the Prehistoric Society* 42, 15–45.

Mercer, R. 1981. *Farming Practice in British Prehistory.* Edinburgh : Edinburgh University Press.

Merrifield R. 1975. *The Archaeology of London.* London: Heinemann Educational.

Merrifield, R. 1983. *London: City of the Romans.* Bath: Pitman Press.

Merriman, N. 1990. *Prehistoric London.* Museum of London: London.

Merriman, N. 2000. 'Changing approaches to the first millennium BC' pp. 35–51 in I. Haynes, H. Sheldon and L. Hannigan (eds.) *London Under Ground: The Archaeology of a City.* Oxford: Oxbow Books.

Miller, P, Saxby, D, and Conheeney, J, 2006. *The Augustine Priory of St Mary Merton, Surrey: Excavations 1976–1990* (Museum of London Archaeology Service Monograph 34). London: Museum of London.

Millett, M. 1990. *The Romanization of Britain: An Essay in Archaeological Interpretation.* Cambridge: Cambridge University Press.

Millett, M. 1994. Evaluating Roman London. *Archaeology Journal* 151, 427–434.

Millett, M. 1996. 'Characterising Roman London' pp. 33–37 in J. Bird, M.W.C Hassel and H. Sheldon (eds.) *Interpreting Roman London: Papers in Memory of Hugh Chapman* (Oxbow Monographs 58). Oxford: Oxbow Books.

Millett, M. 1998. 'Introduction: London as capital?' pp 7-12 in B. Watson (ed.) *Roman London: Recent Archaeological Work* (Journal of Roman Archaeology, Supplementary Series 24). Portsmouth, Rhode Island: Journal of Roman Archaeology.

Milne, G. 1992. *From Basilica to Medieval Market*. London: HMSO.

Milne, G. 1993. *The Port of Roman London* (2nd edition). Batsford / English Heritage.

Milne, G. 1995. *Roman London.* London: Batsford/ English Heritage.

Milne, G. 1996. 'A palace disproved: Reassessing the provincial governor's presence in 1st century London' pp. 49–55 in J. Bird, M. Hassall and H. Sheldon (eds.) *Interpreting Roman London: Papers in Memory of Hugh Chapman* (Oxbow Monograph 58). Oxford: Oxbow Books.

Mjöberg, E. 1905. Über eine schwedische interglaciale Gyrinus-Species. *Geologiska Föreningens i Stockholm Förhandlingar* 27, 233–236.

Mjöberg, E. 1915. Über die Insektenreste der sogennanten "Härnögyttja" im nordlichen Schweden. *Sveriges Geologiska Undersökning*, C268, 1–14.

Moffett, L., Robinson, M.A. & Straker, V. 1989. 'Cereals, fruits and nuts: Charred plant remains from Neolithic sites in England and Wales and the Neolithic economy' pp. 243–261 in A. Milles, D. Williams & N. Gardiner (eds.) *The Beginnings of Agriculture* (British Archaeological Reports, International Series 496). Oxford: British Archaeological Reports.

Moffett, L.C. and Smith, D.N. 1997. Insects and plants from a Late Medieval tenement in Stone, Staffordshire. *Circaea* 12(2), 157–175.

Morris, M. and Smith, D.N. 2008. 'Insect remains' pp. 486–487 in D. Bowsher, T. Dyson, N. Holder and I. Howell (eds) *The London Guildhall: An Achaeological History of a Neighbourhood from Early Mediveal to Modern Times.* (Museum of London Archaeology Service Monograph 36). London: Museum of London.

Mourier, H. Winding, O. and Sunesen, E. 1977. *Collins Guide to Wildlife in House and Home*. London: Collins.

Needham, S. 1991. *Excavation and Salvage at Runnymede Bridge, 1978: The Late Bronze Age Waterfront Site*. London: British Museum.

Oldroyd, H. 1964. *The Natural History of the Flies*. London: British Museum (Natural History).

Ordish, G. 1960. *The Living House.* London: Rupert Hart-Davis.

Osborne, P.J. 1969. An insect fauna of Late Bronze Age date from Wilsford, Wiltshire. *Journal of Animal Ecology* 38, 555–566.

Osborne, P.J. 1971. An insect fauna from the Roman site at Alcester, Warwickshire. *Britannia* 2, 156–165.

Osborne, P.J. 1983. An insect fauna from a modern cesspit and its comparison with probable cesspit assemblages from archaeological sites. *Journal of Archaeological Science* 10, 453–463.

Osborne, P.J. 1988. A late Bronze Age insect fauna from the River Avon, Warwickshire, England: Its implications for the terrestrial and fluvial environment and for climate. *Journal of Archaeological Science* 15, 715–727.

Osborne, P.J. 1989. 'Insects' pp. 96–99 in P. Ashbee, M. Bell & E. Proudfoot (eds.) *Wilsford Shaft. Excavations 1960–62*. London: English Heritage.

Osborne, P.J. 1994. 'Insect remains from pit F and their environmental implications' pp. 217–220 in S. Cracknell & C. Mahany (eds.) *Roman Alcester: Southern Extramural Area. 1964-1966 Excavations, Part 2: Finds and Discussion* (Council for British Archaeology Research Report 97). London: Council for British Archaeology.

Parfitt, S.A. Barendregt, R.W., Breda, M., Candy, I. Collins, M.J.Coope, G.R., Durbidge, P., Field, M.H., Lee, J.R., Lister, A.R. Mutch, R., Penkman, K.E.H. Preece, R.C, Rose, J. Stringer, C.B., Symmons, R. Whittaker, J.R., Wymer, J.J. and Anthony J. Stuart, A. J. 2005. The earliest record of human activity in northern Europe. *Nature* 438, 1008–1012.

Panagiotakopulu, E. 2000. *Archaeology and Entomology in the Eastern Mediterranean: Research into the History of Insect Synanthropy in Greece and Egypt* (British Archaeological Reports, International Series 836). Oxford: British Archaeological Reports.

Panagiotakopulu, E. 2001. New records for ancient pests: Archaeoentomology in Egypt. *Journal of Archaeological Science* 28, 1235–1246.

Panagiotakopulu, E. 2004. Dipterous remains and archaeological interpretation. *Journal of Archaeological Science* 31, 1675–1684

Parker, A.G. and Robinson, M.A. 2003. 'Palaeoenvironmental investigations on the middle Thames at Dorney, UK' pp. 43–60 in A.J. Howard, M.G. Macklin and D.G. Passmore (eds.) *Alluvial Archaeology in Europe.* Lisse: Swets and Zeitlinger.

Parker A.G., Goudie, A.S., Anderson, D.E., Robinson, M.A. and Bonsall, C. 2002. A review of the mid-Holocene elm decline in the British Isles. *Progress in Physical Geography* 26, 1–45.

Perring, D. 1991. *Roman London.* London: Seaby.

Perring, D. and Brigham, T. 2000. 'Londinium and its hinterland: The Roman Period' pp. 120–170 in Museum of London *The Archaeology of Greater London: An Assessment of Archaeological Evidence for Human Presence in the Area now Covered by Greater London.* London: Museum of London Archaeology Service.

Perring, D., Roskams, S. & Allen, P. 1991. *Early Development of Roman London West of the Walbrook* (The Archaeology of Roman London Volume 2/ Council for British Arcaheology Research Report 70). London: Council for British Archaeology.

Perry, D. W., Buckland, P. C. & Snæsdóttir, M. 1985. The Application of numerical techniques to insect assemblages from the site of Stóraborg, Iceland. *Journal of Archaeological Science* 12, 335–345.

Peterken, G. F. 1996. *Natural Woodland: Ecology and Conservation in Northern Temperate Regions.* Cambridge: Cambridge University Press.

Pierce, W. D. 1947. Fossil arthropods of California. 14. A progress report on the McKittrick asphalt field. *Bulletin of the Southern California Academy of Science* 46, 138–143.

Pierce, W. D. 1953. Significance of insect remains in asphalt deposits. *Bulletin of the American Association of Petroleum Geologists* 37, 188–189.

Pipe, A. 1997. 'Animal bone' pp. 231–234 in Thomas, C., Sloane, B. and Philpotts, C (eds.), *Excavations at the Priory and Hospital of St. Mary Spital, London* (Museum of London Archaeology Service Monograph 1). London: Museum of London.

Pipe, A. 2008. 'Invertebrates' p. 217 in in N. Bateman, C. Cowan and R. Wroe-Brown, 2008. *London's Roman Amphitheatre: Guildhall Yard, City of London* (Museum of London Archaeology Service Monograph 10). London: Museum of London.

Pollard, J. 1999. 'These places have their moments: Thoughts on settlement practices in the British Neolithic' pp. 76–93 in J. Brück and M. Goodman (eds.) *Making Places in the Prehistoric World.* London: University College London Press.

Pollard, J. 2000. 'Neolithic occupation practices and social ecologies from Rinyo to Clacton' pp. 363–370 in A. Richie (ed.) *Neolithic Orkney in its European Context.* Cambridge: McDonald Institute.

Pollard, J. 2004. 'A 'movement of becoming': Realms of existence in the early Neolithic of Southern Britain' pp. 55–70 in A.M. Chadwick (ed.) *Stories of the Landscape: Archaeologies of Inhabitation* (British Archaeological Reports, International Series 1238). Oxford: Archaeopress.

Pollard, J. and Reynolds, A. 2002. *Avebury: The Biography of a Landscape.* Stroud: Tempus.

Ponel, P., Orgeas, J., Samways, M.J., Andrieu-Ponel, V., de Beaulieu, J.- L., Reille, M., Roche, P., Tatoni, T. 2003. 110,000 years of Quaternary beetle diversity change. *Biodiversity and Conservation* 12, 2077–2089.

Porter, G. 1997. 'An early Medieval settlement at Guildhall, City of London' pp. 147–152 in G. De Boe and F. Verhaeghe (eds.) *Urbanism in Medieval Europe* (Papers of the 'Medieval Europe Brugge 1997' Conference – Volume 1). Zellik: I.A.P. Rapporten.

Poulton, R. 2004. 'Iron Age Surrey' pp. 51–64 in in J. Cotton, G. Crocker and A. Graham (eds.) *Aspects of Archaeology and History in Surrey.* Guildford: Surrey Archaeological Society.

Preece, R.C. 1999. The mollusca from the last interglacial fluvial deposits if the River Thames at Trafalgar Square, London. *Journal of Quaternary Science* 14, 77–89.

Rackham, J. 1994. 'Economy and environment in Saxon London' pp. 126–135 in J. Rackham (ed.) *Environment and Economy in Anglo-Saxon England* (Council for British Archaeology Research Report 89). York: Council for British Archaeology.

Rackham, J. & Sidell, J. 2000. 'London's landscapes: The changing environment' pp. 11–27 in Museum of London *The Archaeology of Greater London: An Assessment of Archaeological Evidence for Human Presence in the Area now Covered by Greater London.* London: Museum of London Archaeology Service.

Rasmussen, M. 1989. Leaf foddering of livestock in the Neolithic: Archaeobotanical evidence from Weier, Switzerland. *Journal of Danish Archaeology* 8, 51–77.

Reece, R. 2008. 'Satellite, parasite, or just London?' pp. 46–48 in J. Clark, J. Cotton, J. Hall, R. Sherris and H. Swain (eds.) *Londinium and Beyond: Essays on Roman London and its Hinterland for Harry Sheldon* (Council for British Archaeology Research Report 156). London: Council for British Archaeology.

Redfern, R. and Roberts, C. 2005. 'Health in Romano-British urban communities: Reflections from the cemeteries' pp 115–129 in D.N. Smith, M.B. Brickley and W. Smith (eds.) *Fertile Ground: Papers in Honour of Professor Susan Limbrey* (AEA symposia no. 22). Oxford: Oxbow Books.

Reynolds, P.J. 1974. Experimental Iron Age storage pits: An interim report. *Proceedings of the Prehistoric Society* 40, 118–131.

Richards, C. and Thomas, J. 1984. 'Ritual activity and structured deposition in Later Neolithic Wessex' pp. 123–218 in R. Bradley and J. Gardiner (eds.) *Neolithic Studies: A Review of Some Current Research* (British Archaeological Reports, British Series 133). Oxford: British Archaeological Reports.

Richmond, A. 1999. *Preferred Economies* (British Archaeological Reports, British Series 290). Oxford: British Archaeological Reports.

Rielly, K. 2003. 'The animal and fish bone' pp. 315–324 in G. Malcolm, D. Bowsher, and R. Cowie, 2003. *Middle Saxon London: Excavations at the Royal Opera House 1989-99* (Museum of London Archaeology Service Monograph 15). London: Museum of London.

Rielly, K. 2006 'The animal bone' pp. 130–141 in Seeley, D., Phillpotts, C. and Samuel, M. (eds.) *Winchester Palace: Excavations at the Southwark Residence of the Bishops of Winchester* (Museum of London Archaeology Service Monograph 31). London: Museum of London.

Robinson, M. A. 1978. 'A comparison between the effects of man on the environment of the first gravel terrace and flood-plain of the Upper Thames Valley during the Iron Age and Roman periods' pp. 35–43 in S. Limbrey & J.G. Evans (eds.) *The Effect of Man on the Landscape: The Lowland Zone* (Council for British Archaeology Research Report 21). London: Council for British Archaeology.

Robinson, M. A. 1979. 'The biological evidence' pp 77–133 in G. Lambrick & M. A. Robinson (eds.) *Iron Age and Roman Riverside Settlements at Farmoor, Oxfordshire* (Council for British Archaeology Research Report 32). London: Council for British Archaeology.

Robinson, M.A. 1981. 'The use of ecological groupings of Coleoptera for comparing sites' pp. 251–86 in M. Jones and G. Dimbleby (eds.) *The Environment of Man: The Iron Age to the Anglo-Saxon Period* (British Archaeological Reports, British Series 87). Oxford: British Archaeological Reports.

Robinson, M.A. 1983. 'Arable/ pastoral ratios from insects?' pp. 19–47 in M. Jones (ed.) *Integrating the Subsistance Economy* (British Archaeological Reports, International Series 181). Oxford: British Archaeological Reports.

Robinson, M. A. 1991. 'The Neolithic and Late Bronze Age insect assemblages' pp. 277–325 in S. Needham (ed.) *Excavation and Salvage at Runnymede Bridge, 1978: The Late Bronze Age Waterfront Site.* London: British Museum.

Robinson, M. 1993. 'The Iron Age environmental evidence' pp. 101–20 in T. G. Allen and M. A. Robinson (eds.) *The Prehistoric Landscape and Iron Age Enclosed Settlement at Mingies Ditch, Hardwick-with-Yelford,Oxon* (Thames Valley Landscapes: The Windrush Valley Volume 2). Oxford: Oxford Archaeological Unit, Oxford.

Robinson, M. A. 2000a 'Middle Mesolithic to Late Bronze Age insect assemblages and an Early Neolithic assemblage of waterlogged macroscopic plant remains' pp 146–167 in S. P. Needham (ed) *The Passage of the Thames. Holocene Environment and Settlement at Runnymede.* London: British Museum Press.

Robinson, M. A. 2000b 'Coleopteran evidence for the elm decline, Neolithic activity in woodland, clearance and the use of the landscape' pp 27–36 in A. S. Fairbairn (ed.) *Plants in Neolithic Britain and Beyond.* (Neolithic Studies Group Seminar Papers 5). Oxford: Oxbow Books.

Robinson, M.A. 2000c 'Further considerations of Neolithic charred cereals, fruits and nuts' pp. 85–90 in A. S. Fairbairn (ed.) *Plants in Neolithic Britain and Beyond* (Neolithic Studies Group Seminar Papers 5) Oxford: Oxbow Books.

Robinson, M. & Lambrick, G. 1979. *Iron Age and Roman Riverside Settlements at Farmoor, Oxfordshire* (Council for British Archaeology Research Report 32). London: Council for British Archaeology.

Rousseau, M. 2011. Paraffin flotation for archaeoentomological research: Is it really efficient? *Environmental Archaeology* 16, 58–64.

Rowley-Conwy, P. 1981. 'Slash and burn in the temperate European Neolithic' pp 85–96 in R. Mercer (ed.) *Farming Practice in British Prehistory*. Edinburgh: Edinburgh University Press.

Rowley-Conwy, P. 1984. 'Dung, dirt and deposits: Site formation under conditions of near-perfect preservation at Qasr Ibrim, Egyptian Nubia' pp. 25-32 in R. Luff and P. Rowley-Conwy (eds.) *Whither Environmental Archaeology?* Oxford: Oxbow Books.

Rowsome, P. 1998. 'The development of the Town Plan of early Roman London' pp. 35–46 in B. Watson (ed.) *Roman London: Recent Archaeological Work* (Journal of Roman Archaeology, Supplementary Series 24). Portsmouth, Rhode Island: Journal of Roman Archaeology.

Rowsome, P. 2000. *Heart of the City: Roman, Medieval and Modern London Revealed by Archaeology at 1 Poultry*. London: Museum of London

Rowsome, P. 2008. 'Mapping Roman London: Identifying its urban patterns and interpreting their meaning' pp. 25–32 in J. Clark, J. Cotton, J. Hall, R. Sherris and H. Swain (eds.) *Londinium and Beyond: Essays on Roman London and its Hinterland for Harry Sheldon* (Council for British Archaeology Reserch Report 156). London: Council for British Archaeology.

Rowsome, P and Treviel, P. 1988. No 1, Poultry. *Current Archaeology* 158, 50–56.

Rousell, A. 1943. 'Stöng, Þjórrsádalur' pp. 72–97 in M. Stenberger and E. Munksgaard (eds.) *Forntida Gårdar I Island*. Copenhagen: Munksgaard.

Ruiz, Z., Brown, A.G. & Langdon, P.G. 2006. The potential of chironomid (Insecta: Diptera) larvae in archaeological investigations of floodplain and lake settlements. *Journal of Archaeological Science* 33, 14–33.

Salmond, K.F. 1957. The insect and mite fauna of a Scottish flour mill. *Bulletin of Entomological Research* 47, 621–630.

Scaife, R. 2000. 'Palynology and palaeoenvironments' pp. 168–187 in S.P. Needham (ed.) *The Passage of the Thames: Holocene Environment and Settlement at Runnymede* (Runnymede Bridge and Research Excavations Volume 1). London: British Museum Press.

Scarre, C. (ed.) 2002. *Monuments and Landscape in Atlantic Europe*. London: Routledge.

Schelvis, J. 2000. 'Remains of Mites (Acari) from the Iron Age site' pp. 272–276 in M. Bell, A. Caseldine and H. Neumann (eds.) *Prehistoric Intertidal Archaeology in the Welsh Severn Estuary* (Council for British Archaeology Research Report 120). London: Council for British Archaeology.

Schofield, J. 2000. 'London: Buildings and Defences 1200–1600' pp. 223–238 in I. Haynes, H. Sheldon and L. Hannigan (eds.) *London Underground: The Archaeology of a City*. Oxford: Oxbow Books.

Scudder, S. H. 1900. Canadian Fossil Insects, 4: Additions to the Coleopterous fauna of the interglacial clays of the Toronto district, pp. 67-92 in Dawson, G. (ed.) *Geological Survey of Canada, Contributions to Canadian Palaeontology 2*. Ottawa: S.E. Dawson.

Seeley, F. and Drummond-Murray, J. 2005. *Roman Pottery Production in the Walbrook Valley: Excavations at 20–28 Moorgate, City of London, 1998–2000* (Museum of London Arcaheology Service Monograph Series 25). London: Museum of London.

Seeley, F., Phillpotts, C. and Samuel, M. 2006 (eds.) *Winchester Palace: Excavations at the Southwark Residence of the Bishops of Winchester* (Museum of London Archaeology Service Monograph 31). London: Museum of London.

Shanks, M. and Tilley, C. 1982. 'Ideology, symbolic power and ritual communication: A reinterpretation of Neolithic mortuary practices' pp. 129–154 in I. Hodder (ed.) *Symbolic and Structural Archaeology*. Cambridge: Cambridge University Press.

Shanks, M. and Tilley, C. 1987. *Social Theory and Archaeology*. London: Polity Press.

Sheldon, H. 2000. 'Roman Southwark' pp. 121–151 in I. Haynes, H. Sheldon and L. Hannigan (eds.) *London Underground: The Archaeology of a City*. Oxford: Oxbow Books.

Shotton, F.W. and Osborne P.J. 1965. The fauna of the Hoxnian interglacial deposits of Nechells, Birmingham. *Philosophical Transactions of the Royal Society* B 248, 353–378.

Sidell, J. 2000. 'Twenty-five years of environmental archaeology in London' pp. 284–294 in I. Haynes, H. Sheldon and L. Hannigan (eds.) *London Underground: The Archaeology of a City*. Oxford: Oxbow Books.

Sidell, J. 2008 'Londinium's Landscape' pp. 62–68 in in J. Clark, J. Cotton, J. Hall, R. Sherris and H. Swain (eds.) *Londinium and Beyond: Essays on Roman London and its Hinterland for Harry Sheldon* (Council for British Archaeology Reserch Report 156). London: Council for British Archaeology.

Sidell, J., Williams, K., Scaife, R. and Cameron, N. 2000. *The Holocene Evolution of the London Thames* (Museum of London Archaeology Service Monograph 5). London: Museum of London.

Sidell, J., Cotton, J., Rayner, L. and Wheeler, L. 2002. *The Prehistory and Topography of Southwark and Lambeth* (Museum of London Archaeology Service Monograph 14). London: Museum of London.

Simmonds, I.G. 1975. Towards an ecology of Mesolithic man in the uplands of Great Britain. *Journal of Archaeological Science* 2, 1–15.

Simmonds, I.G. 1996. *The Environmental Impact of Later Mesolithic Cultures*. Edinburgh: Edinburgh University Press.

Simmonds, I. G. and Dimbleby, G.W. 1974. The possible role of ivy in the Mesolithic economy of Western Europe. *Journal of Archaeological Science* 1, 291–296.

Simpson, T. 2001. The Roman well at Piddington, Northhamptonshire, England: an investigation of the insect fauna. *Environmental Archaeology* 6, 91–96.

Skidmore, P. 1999. 'The Diptera' pp. 341–343 in A. Connor and R. Buckley *Roman and Medieval Occupation in Causeway Lane, Leicester* (Leicester Archaeological Monographs 5). Leicester: Leicester University Press.

Sloane, B. and Malcolm, G. 2004. *Excavations at the Priory of the Order of the Hospital of St John of Jerusalem, Clerkenwell, London* (Museum of London Archaeology Service Monograph 20). London: Museum of London.

Solomon, M.E. and Adamson, B.E. 1956. The powers of survival of storage and domestic pests under winter conditions in Britain. *Bulletin of Entomological Research* 46, 311–355.

Smith, D. 1996a. Thatch, turves and floor deposits: A survey of Coleoptera in materials from abandoned Hebridean blackhouses and the implications for their visibility in the archaeological record. *Journal of Archaeological Science* 23, 161–174.

Smith, D. 1996b. 'Hebridean blackhouses and a speculative history of the 'culture favoured' Coleoptera of the Hebrides' pp. 207–217 in D. Gilbertson, M. Kent and J. Grattan (eds.) *The Outer Hebrides: The Last 14,000 Years* (Sheffield Environmental and Archaeological Research Campaign in the Hebrides. Volume 2). Sheffield: Sheffield Academic Press.

Smith, D. 1997a. *The insect remains from Mancetter Mill Lane, Roman well.* (Unpublished report prepared for Keith Scot, director of excavations at Mancetter).

Smith, D. 1997b. 'The insect fauna' pp. 245–47, in C. Thomas, B. Sloane and C. Philpotts (eds.), *Excavations at the Priory and Hospital of St. Mary Spital, London* (Museum of London Archaeology Service Monograph 1). London: Museum of London.

Smith, D. 1998. Beyond the barn beetles: Difficulties in using some Coleoptera as indicators for stored fodder. *The Journal of Environmental Archaeology* 1, 63–70.

Smith, D.N. 1999. *Atlas Wharf, Isle of Dogs: Paleoentomological analysis.* (The University of Birmingham Environmental Archaeology Services. Report 8). Unpublished report to Museum of London Archaeology Service.

Smith, D. N. 2000a. 'Detecting the nature of materials on farms using Coleoptera' pp. 71–84 in J. P. Huntley & S. Stallibrass (eds.) *Taphonomy and Interpretation.* (Association for Environmental Archaeology Symposia No. 14). Oxford: Oxbow Books.

Smith, D. 2000b. 'Disappearance of elmid "riffle beetles" from lowland river systems - the impact of alluviation' pp. 75–80 in T. O'Connor & R. Nicholson, (eds.), *People as an Agent of Environmental Change* (Symposia of the Association for Environmental Archaeology 16). Oxford: Oxbow Books.

Smith, D.N. 2002. 'Insect remains' pp. 113–115 in B. Barber & C. Thomas C. (eds.) *The London Charterhouse.* (Museum of London Archaeology Service Monograph 10). London: Museum of London.

Smith, D. N. 2004. 'The insect remains from the well' pp. 81–88 in M. C. Bishop (ed.) *Inveresk Gate: Excavations in the Roman Civil Settlement at Inveresk, East Lothian, 1996–2000.* (Scotish Trust for Archaeological Research Monograph 7). Loanhead, Midlothian.

Smith, D.N. 2006 'The insect remains' pp. 142–144 in D. Seeley, C. Phillpotts and M. Samuel (eds.) *Winchester Palace: Excavations at the Southwark Residence of the Bishops of Winchester* (Museum of London Archaeology Services Monograph 31). London: Museum of London.

Smith, D.N. 2010. 'The insect remains' pp. 921–925 and pp. 1481–1489 on CD in C. Howard-Davis (ed.) *The Carlisle Millennium Project: Excavations in Carlisle 1998–2001. Volume 2: The Finds* (Lancaster Imprints 15). Lancaster: Oxford Archaeology North.

Smith, D.N. 2011a. 'Insects from Northfleet' pp. 88–90 in C. Barnett, J. J. McKinley, E. Stafford, J. M. Grimm and C. J. Stevens (eds.) *Settling the Ebbsfleet Valley: High Speed I Excavations at Springhead and Northfleet, Kent, The Late Iron Age, Roman, Saxon and Medieval Landscape: Volume 3: Late Iron Age to Roman Human Remains and Environmental Reports*. Oxford/ Salisbury: Oxford-Wessex Archaeology.

Smith, D.N. 2011b. 'The insect remains' p. 342 and contributions to text and CD in M. Burch and P. Treveil, with D. Keene (eds.) *The Development of Early Medieval and Later Poultry and Cheapside: Excavations at 1 Poultry and Vicinity, City of London* (Museum of London Archaeology Services Monograph Series 38). London: Museum of London.

Smith, D.N. 2012 'The insect remains' pp. 559–563 in J. Hill and P. Rowsome (eds.) *Roman London and the Walbrook Stream Crossing: Excavations at 1 Poultry and Vicinity, City of London* (Museum of London Archaelogy Monograph Series 37). London: Museum of London Archaeology.

Smith, D.N. and Chandler, G. 2004. 'Insect remains, pp. 389–94 in B. Sloane and G. Malcolm (eds.) *Excavations at the Priory of the Order of the Hospital of St John of Jerusalem, Clerkenwell, London* (Museum of London Archaeology Service Monograph 20). London: Museum of London.

Smith, D.N. & Howard, A.J. 2004. Identifying changing fluvial conditions in low gradient alluvial archaeological landscapes: Can Coleoptera provide insights into changing discharge rates and floodplain evolution? *Journal of Archaeological Science* 31, 109–20.

Smith, D.N. and Kenward, H.K. 2011. Roman grain pests in Britain: Implications for grain supply and agricultural production . *Britannia* 42, 243–262.

Smith, D.N. and Kenward, H.K. Forthcoming 2013. 'Well, Sextus what can we do with this?': The disposal and use of insect-infested grain in Roman Northern Europe. *Environmental Archaeology.*

Smith, D., Letts, J. P. and Cox, A. 1999. Coleoptera from Late Medieval smoke blackened thatch (SBT): Its archaeological implications. *Environmental Archaeology* 4, 9–18.

Smith, D.N. Letts, J. and Jones, M. 2005. The insects from non-cereal stalk smoked blackened thatch. *Environmental Archaeology* 10, 171–178.

Smith, D.N. and Morris, M. 2008. 'Insect remains' pp. 218–219 in N. Bateman, C. Cowan and R. Wroe-Brown (eds.) *London's Roman Amphitheatre: Guildhall Yard, City of London* (Museum of London Archaeology Service Monograph 10). London: Museum of London.

Smith, D., Osborne, P. and Barratt, J. 2000. 'Beetles and evidence of past environments at Goldcliff' pp. 245–60 in M, Bell, A. Caseldine and H. Neumann (eds.), *Prehistoric Intertidal Archaeology in the Welsh Severn Estuary.* (Council for British Archaeology Research Report 120). London: Council for British Archaeology.

Smith, D.N., Roseff, R., Bevan, L., Brown, A.G. Butler, S, G. Hughes, A. Monckton. 2005. Archaeological and environmental investigations of a Late Glacial and Holocene river valley sequence on the River Soar, at Croft, Leicestershire. *The Holocene* 15, 353–377.

Smith, D.N. & Whitehouse, N. 2005. 'Not seeing the trees for the woods: A palaeoentomological perspective on Holocene woodland composition' pp. 136–161 in D.N. Smith, M.B. Brickley and W. Smith (eds.) *Fertile Ground: Papers in Honour of Professor Susan Limbrey* (Association for Environmental Archaeology Symposia No. 22). Oxford: Oxbow Books.

Smith K.G.V. 1973. *Insects and Other Arthropods of Medical Importance.* London: British Museum (Natural History).

Smith, D.N., Whitehouse, N, Bunting, M.J. and Chapman, H. 2010b. Can we characterise 'openness' in the Holocene palaeoenvironmental record? Analogue studies from Dunham Massey deer park and Epping Forest, England. *The Holocene.* 20, 215–229.

Smith K.G.V. 1989. *An Introduction to the Immature Stages of British Flies.* (Handbooks for the Identification of British Insects Vol. 10 part 14). London: Royal Entomological Society of London.

Strobel, P. & Pigorini, L. 1864. Le terremare e le palafitte del Parmense, seconda relazione. *Atti della Societa italiana di Scienze Naturali* (Milan) 7, 36–37.

ter Braak, C. J.F. 1987. Canonical correspondance analysis: A new eigenvector method for multivariate direct gradient analysis. *Vegetatio* 69, 69–77.

ter Braak, C. J.F. and Smilauer, P. 2002. *CANOCO Reference Manual and CanoDraw for Widows User's Guide: Software for Canonical Community Ordination (Version 4.5).* Ithaca, New York: Microcomputer Power.

Tetlow, E. 2004. *The Archaeoentomology of the Salt Marshes and Woodland of the Severn Estuary.* Unpublished PhD Thesis, Institute of Archaeology and Antiquity, The University of Birmingham.

Thomas, C. and Rackham, D.J. 1996. Bramcote Green, Bermondsey: A Bronze Age trackway and palaeoenvironmental sequence. *Proceedings of the Prehistoric Society* 61, 221–253.

Thomas, C., Slone, B. and Phillpotts, C. 1997. *Excavations at the Priory and Hospital of St. Mary Spital, London.* (Museum of London Archaeology Service Monograph 1). London: Museum of London.

Thomas, J. 1990. Silent running: The ills of environmental archaeology. *Scottish Archaeological Review* 7, 2–7.

Thomas, J. 1991. *Rethinking the Neolithic.* Cambridge: Cambridge University Press.

Thomas, J. 1999. *Understanding the Neolithic.* London: Routledge.

Thomas, J. 2003. Thoughts on the 'repacked' Neolithic Revolution. *Antiquity* 77, 67–74.

Thomas, J. 2004. Recent debates on the Mesolithic – Neolithic transition in Britain and Ireland. *Documenta Praehistorica* 31, 113–130.

Thomas, J. 2008. 'The Mesolithic – Neolithic transition in Britian' pp. 58–89 in J. Pollard (ed.) *Prehistoric Britain.* Oxford: Blackwell Publishing.

Thomas, R. 1999. 'Rise and fall: The deposition of Bronze age weapons in the Thames Valley and the Fenland' pp 116–122 in A.F. Harding (ed.) *Experiment and Design: Archaeological Studies in Honour of John Coles.* Oxford: Oxbow Books.

Tilley, C. 1994. *A Phenomenology of Landscape: Places, Paths and Monuments.* Oxford: Berg.

Tilley, C. 2003. *The Materiality of Stone: Explorations in Landscape Phenomenology.* Oxford: Berg.

Tipping, R. 1996. 'Microscopic charcoal records, inferred human activity and climate change in the Mesolithic of northernmost Scotland' pp. 29–61 in T. Pollard and A. Morrison (eds.) *The Early Prehistory of Scotland.* Edinburgh: Edinburgh University Press.

Tipping, R. 2004. 'Interpretative issues concerning the driving forces of vegetation change in the Early Holocene of the British Isles' pp. 45–54 in A. Saville (ed.) *Mesolithic Scotland and its Neighbours.* Edinburgh: Society of Antiquaries of Scotland.

Troels-Smith, J. 1960. Ivy, mistletoe and elm. Climatic indicators – fodder plants. A contribution to the interpretation of the pollen zone boarder VII-VIII. *Danmarks Geologiske Undersøglese II.* (Raekke 4/4), 1–32.

Tyers, I. 2008. 'A gazetteer of tree-ring dates from Roman London' pp. 69–74 in in J. Clark, J. Cotton, J. Hall, R. Sherris and H. Swain (eds.) *Londinium and Beyond: Essays on Roman London and its Hinterland for Harry Sheldon* (Council for British Archaeology Research Report 156). London: Council for British Archaeology.

van de Noort, R. and O'Sullivan, A. 2006. *Rethinking Wetland Archaeology.* London: Duckworth.

van Der Veen, M. 1992. *Crop Husbandry Regimes: An Archaeolobotanical Study of Farming in Northern England.* Sheffield: Sheffield University Press.

Vera, F.W.M. 2000. *Grazing Ecology and Forest History.* Oxon: CABI Publishing.

Vince, A. 1990. *Saxon London: An Archaeological Investigation.* London: Seaby.

Vince, A. 2000. 'The study of Medieval Pottery in London' pp. 239–251 in I. Haynes, H. Sheldon and L. Hannigan (eds.) *London Underground: The Archaeology of a City.* Oxford: Oxbow Books.

Wait, G. and Cotton, J. 2000. 'The Iron age' pp. 101–118 in Museum of London *The Archaeology of Greater London: An Assessment of Archaeological Evidence for Human Presence in the Area now Covered by Greater London.* London: Museum of London.

Walker, I.R. & Cwynar L.C. 2006. Midges and palaeotemperature reconstruction—the North American Experience. *Quaternary Science Reviews* 25, 1911–1925.

Wheeler, R.E.M. 1934. The topography of Saxon London. *Antiquity* 8, 290–302.

Whitehouse, N.J. 1997. Insect faunas associated with *Pinus sylvestris* L. from the mid-Holocene of the Humberhead Levels, Yorkshire, U.K. *Quaternary Proceedings* 5, 293–303.

Whitehouse, N.J. 2000. Forest fires and insects: Palaeoentomological research from a sub-fossil burnt forest. *Palaeogeography, Palaeoclimatology, Palaeoecology* 164, 231–246.

Whitehouse N. J. 2004. Mire ontogeny, environmental and climate change inferred from fossil beetle successions from Hatfield Moors, eastern England. *The Holocene* 14, 79–93.

Whitehouse, N.J. 2006. The Holocene British and Irish ancient forest fossil beetle fauna: Implications for forest history, biodiversity and faunal colonisation. *Quaternary Science Reviews* 25, 1755–1789.

Whitehouse, N.J. and Smith, D.N. 2004 'Islands' in Holocene forests: Implications for forest openness, landscape clearance and 'culture steppe' species. *Environmental Archaeology* 9, 203–12.

Whitehouse, N. and Smith D. 2010. What is "natural"? Forest composition, open-ness and the British "wildwood": Implications from palaeoentomology for Holocene development and landscape structure. *Quaternary Science Reviews* 29, 539–553.

Whittle, A. 1997. 'Moving on and moving around: Neolithic settlement mobility' pp. 15-22 in P. Topping (ed.) *Neolithic Landscapes* (Neolithic Studies Group Seminar Papers 2). Oxford: Oxbow Books.

Whittle, A. 2000. 'Bringing plants into the taskscape' pp. 1–8 in A. S. Fairbairn (ed.) *Plants in Neolithic Britain and Beyond* (Neolithic Studies Group Seminar Papers 5) Oxford: Oxbow Books.

Whittle, A. 2002. 'Conclusion: Long conversations, concerning time, descent and place in the world' pp. 192–204 in C. Scarre (ed.) *Monuments and Landscape in Atlantic Europe*. London: Routledge.

Whittle, A. and Pollard, J. 1999. 'A harmony of symbols: Wider meanings' pp. 381–390 in A. Whittle, J. Pollard and C. Grigson (eds.) *The Harmony of Symbols: The Windmill Hill Causeway Enclosure.* Oxford: Oxbow Books.

Whytehead, R. 1988. 'The excavation at Jubilee Hall' pp. 49–66 in R. Cowie and R.L. Whytehead (eds.) Two Middle Saxon occupation sites: excavations at Jubilee Hall and 21-22 Maiden Lane, WC1. *Transactions of the London and Middlesex Archaeology Society* 39, 47–163.

Wilkinson, K.N. and Stevens, C. 2003. *Environmental Archaeology: Approaches, Techniques and Applications.* Stroud: Tempus Publishing.

Wilkinson, K.N., Scaife, R.G. and Sidell, E.J. 2000. Environmental and sea level changes in London from 10,5000 BP to the present: A case study from Slivertown. *Proceedings of the Geologists Association* 111, 41–54.

Wilkinson, T.J. and Murphy, P.L. 1986. Archaeological survey of an intertidal zone: The submerged landscape of the Essex coast, England. *Journal of Field Archaeology* 13, 177–94.

Wilkinson, T.J. and Murphy, P.L. 1995. *The Archaeology of the Essex Coast. Volume 1: The Hullbridge Survey* (East Anglian Archaeology 71). Chelmsford: Essex County Council Archaeology Section.

Woodward, P and Woodward, A. 2004. Dedicating the town: Urban foundation deposits in Roman Britain. *World Archaeology* 36, 68–86.

Wroe-Brown, R. 1988. Bull Wharf: Queenhithe. *Current Archaeology* 158, 75–77.

Wylie, J.A. and Stubbs, H.W. 1983. The plague of Athens, 430–428 B.C.: Epidemic and epizootic. *Classical Quarterly* 33, 6–11.

Yates, D.T. 1999. Bronze Age field systems in the Thames Valley. *Oxford Journal of Archaeology* 18, 157–170.

Yates, D.T. 2001. 'Bronze Age agricultural intensification in the Thames Valley and Estuary' pp. 65–82 in J. Brück (ed.) *Bronze Age Landscapes: Transactions and Transformation.* Oxford: Oxbow Books.

Yule, B. 2005. *A Prestigious Roman Building Complex on the Southwark Waterfront: Excavations at Winchester Palace, London 1983-1990* (Museum of London Archaeology Service Monograph Series 23). London: Museum of London.

Zvelebil, M. 1994. Plant use in the Mesolithic and its role in the transformation to farming. *Proceedings of the Prehistoric Society* 60, 35–74.

INDEX